AF615656

THERMAL MACHINING PROCESSES

Published by:

Society of Manufacturing Engineers
Marketing Services Dept.
One SME Drive
P.O. Box 930
Dearborn, Michigan 48128

THERMAL MACHINING PROCESSES

First Edition

Library of Congress Catalog Card Number: 79-62917

International Standard Book Number: 0-87263-049-8

Manufactured in the United States of America

SME wishes to express its acknowledgement and appreciation to the following publications for supplying the various articles reprinted within its contents

Manufacturing Engineering
Society of Manufacturing Engineers
One SME Drive
P.O. Box 930
Dearborn, Michigan 48128

Metal Stamping
American Metal Stamping Association
27027 Chardon Road
Richmond Heights, Ohio 44143

Modern Machine Shop
Gardner Publications
600 Main Street
Cincinnati, Ohio 45202

Production Engineering
Penton/IPC Publications
Penton Plaza
Cleveland, Ohio 44114

European Plastic News
IPC Business Press Ltd.
33-39 Bowling Green Lane
London, ECI
England

Laser Focus
385 Elliot Street
Newton, Massachusetts 02164

Machine and Tool Blue Book
Hitchcock Publishing Company
Hitchcock Building
Wheaton, Illinois 60187

Production
Bramson Publishing Company
Box 101
Bloomfield Hills, Michigan 48013

Tooling and Production
Huebner Publications, Inc.
5821 Harper Road
Solon, Ohio 44139

Metal Progress
American Society for Metals
Metals Park, Ohio 44073

Nontraditional Machining Guide
Machinability Data Center/Metcut
Research Associates Inc.
3980 Rosslyn Drive
Cincinnati, Ohio 45209

Grateful acknowledgement also to:

The British Library Board
Boston Spa
Wetherby
W. Yorks LS23 7BQ England

PREFACE

Thermal machining was introduced to industry in the middle 1940's by Russian scientists who attempted to prolong the life of contacts used in electrical circuits. While their particular work was not successful, they did conclude that productive machining could be accomplished by controlling the spark in an oil dielectric.

About the same time, an engineer in the United States, Victor Harding investigated spark discharge usage to remove small, broken taps and drills in expensive aircraft components. Mr. Harding's early research was the basis for the development of salvage-type EDM machines and then the precision, controlled-pulse EDM machines which followed.

With the development of highly dependable solid-state electronics, other forms of thermal machining became practical. Today, laser machining is now being phased into industry. Plasma machining is assisting in higher metal removal rates. Electron beam machining is being further developed each day.

In all cases, thermal machining requires capable engineering in both the electronics and the mechanical fields. We are currently in the process of major changes in designing machine tools. Early machine tools were the product of mechanical design engineers developing a product which suited industrial needs. Now that electronics have been accepted into thermal machining, electronic engineers perform a major portion of machining development.

This book is intended to introduce the reader to thermal machining processes. Basic information is presented on each of the methods. In this way, the reader can use the information to decide how and when thermal machining can be used in manufacturing operations. This volume should supplement the data which should be obtained by the user before incorporating the required process into the machining operations.

This book, as part of the SME Manufacturing Update Series, presents the most up-to-date and practical information available. The authors are experts in their field. (Many of the entries include the author's name. Author's titles are those that they held when they wrote the journal article or technical paper.) Grateful acknowledgement is due to each of those authors for his contribution.

I also wish to express my gratitude to the publications who supplied some of the material in this volume. These include: ***Manufacturing Engineering, Metal Stamping, Modern Machine Shop, Production Engineering, European Plastic News, Laser Focus, Machine and Tool Blue Book, Production, Tooling and Production,*** and ***Metal Progress.*** My thanks also to the members of the SME Marketing Services Department for their efforts in producing this book.

E. C. Jameson
Director of Marketing
Raycon Corporation
Technical Advisor

SME

The informative volumes of the Manufacturing Update Series are part of the Society of Manufacturing Engineers' effort to keep its members better informed on the latest trends and developments in engineering.

With 50,000 members, SME provides a common ground for engineers and managers to share ideas, information and accomplishments.

An overwhelming mass of available information requires engineers to be concerned about keeping up-to-date, in other words, continuing education. An SME Member can take advantage of numerous opportunities, in addition to the books of the Manufacturing Update Series, to fulfill his continuing educational goals. These opportunities include:

- Chapter programs through the over 200 chapters which provide SME members with a foundation for involvement and participation.
- Educational programs including seminars, clinics, programmed learning courses and videotapes.
- Conferences and expositions which enable engineers to see, compare, and consider the newest manufacturing equipment and technology.
- Publications including Manufacturing Engineering, the SME Newsletter, Technical Digest and a wide variety of books.
- SME's Manufacturing Engineering Certification Institute formally recognizes manufacturing engineers and technologists for their technical expertise and knowledge acquired through years of experience.

In addition, the society works continuously with the American National Standards Institute, the International Standards Organization and other organizations to establish the highest possible standards in the field.

SME members have discovered that their membership broadens their knowledge throughout their career.

In a very real sense, it makes SME the leader in disseminating and publishing technical information for the manufacturing engineer.

TABLE OF CONTENTS

CHAPTER 1

BASICS

Reprinted from the Nontraditional Machining Guide

Thermal Nontraditional Machining

By Guy Bellows

Machining Data Analyst
Machinability Data Center
Metcut Research Associates Inc.

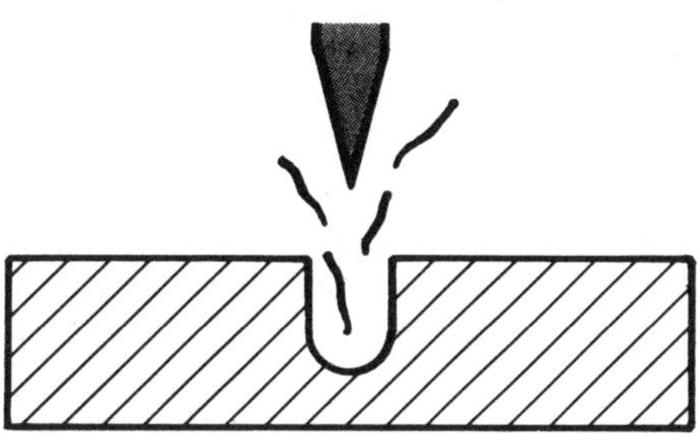

Very few of the thermal material removal processes use the term *thermal* in their name! Laser, discharge, plasma, or beam are typical of the descriptions used. Upon analysis, however, all use high temperatures to melt or vaporize material from the workpiece. The principal energy mode is thermal with the electrical discharge, light beam or electron beam acting as the source of concentrated heat.

Thermal Nontraditional Machining Processes

EBM	-	Electron Beam Machining
EDG	-	Electrical Discharge Grinding
EDM	-	Electrical Discharge Machining
EDS	-	Electrical Discharge Sawing
EDWC	-	Electrical Discharge Wire Cutting
LBM	-	Laser Beam Machining
LBT	-	Laser Beam Torch
PBM	-	Plasma Beam Machining

Electrical discharge machining is the oldest of the nontraditional material removal processes. It had its first real production activity during World War II when it was used to shape workpieces from war generated high strength materials. It has come a long way from the original sparking "tap buster". EDM machines are so well developed and used in so many ways that the process is an "old friend" to the extent that many now consider it to be a conventional process. The application of computer numerical control, tracing heads and electronic controls to manipulate the discharge keeps EDM in the forefront of the commercially usable NTM processes.

The laser, electron beam and plasma beam machining techniques all make use of the ease and versatility of electricity to control these basically thermal material removal processes. This ease of control also provides opportunities for adaptive control, automation and integration into transfer line systems.

The surface integrity effects resulting from the thermal NTM processes are summarized in Table V.

ELECTRICAL DISCHARGE GRINDING

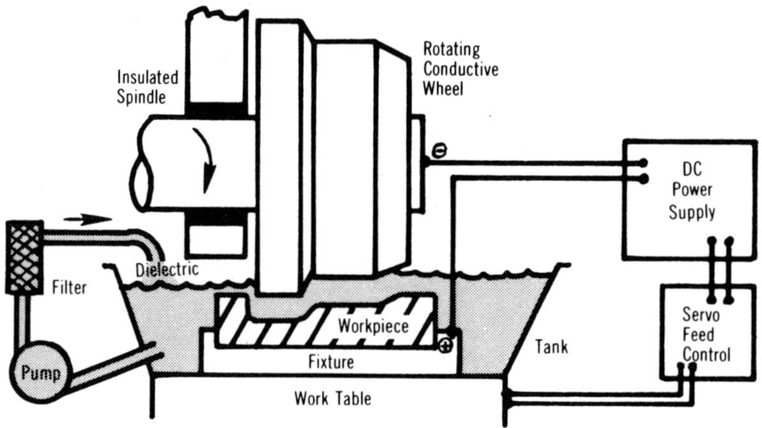

Electrical discharge grinding (EDG) is the removal of a conductive material by rapid repetitive spark discharges between a rotating tool and the workpiece, which are separated by a flowing dielectric fluid. The spark gap is servo controlled with a setting between 0.0005 and 0.003 inch. The d-c power source has capabilities ranging from 30 to 100 volts, 1/2 to 200 amperes and 2 to 500 kilohertz. The conductive wheel, usually graphite, rotates at 100 to 600 sfpm in a dielectric bath of filtered hydrocarbon oil. The insulated wheel and work table are connected to the d-c pulse generator with positive on the workpiece being "standard". Higher currents produce faster cutting, rougher finishes and deeper heat-affected zones in the workpiece. Wheel wear ranges from 100:1 to 0.1:1 with an average of 3:1, depending upon current density, work material, wheel material, dielectric and sharpness of corner details.

PRACTICAL APPLICATIONS

Greater accuracy in cutting hard materials such as form tools or tungsten carbide throwaway bits is possible with EDG even though its cutting rates are low. Lamination die grinding in the hardened state is a frequent use of EDG. The absence of significant cutting forces permits grinding fragile shapes in any conductive material.

MATERIAL REMOVAL RATES AND TOLERANCES

Material removal rates range from 0.01 to 0.15 cubic inch per hour with the higher figures accompanied by finishes in the 63- to 125-microinch AA range. Corner radius is dependent upon overcut values used and ranges from 0.0005 to 0.005 inch. Tolerances to ±0.0002 inch are normal with ±0.000050 inch achievable. Finishes improve with an increase of spark frequency and are typically 16 to 32 microinches AA. The melting, vaporizing and resolidification of the surface of the workpiece leaves a heat-affected zone that can be from a few ten thousandths to a few thousandths of an inch deep. Hardness alterations occur which also affect the material properties. Highly stressed applications should

ELECTRICAL DISCHARGE GRINDING

have these affected layers removed or modified to insure the best surface integrity of the component.

AVAILABILITY

Equipment is regularly available in a wide range of sizes for EDG.

Wheel and fixture for EDG of residual stress specimens. Dielectric tank was drained for picture.

ELECTRICAL DISCHARGE MACHINING

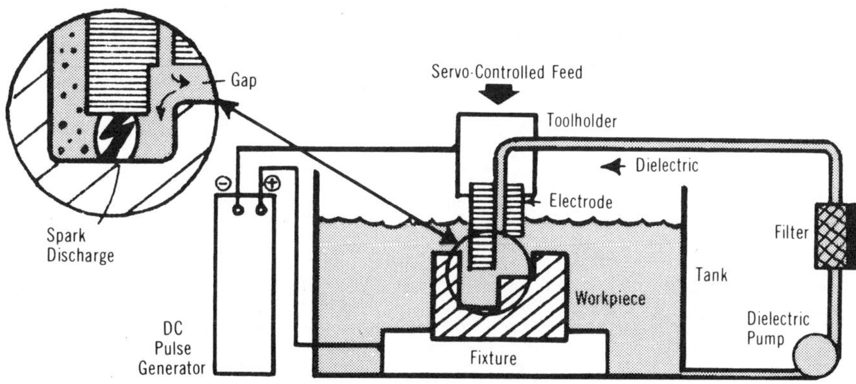

Electrical discharge machining (EDM) removes electrically conductive material with rapid repetitive spark discharges from a pulsating d-c power supply with a dielectric flowing between the workpiece and the tool. The shaped tool (electrode) is fed into the workpiece under servo control until a spark discharge breaks down the dielectric fluid. The frequency (200 to 500,000 sparks per second) and energy per spark are set and controlled with a d-c power source. The servo control maintains a constant gap between the tool and the workpiece while advancing the electrode. The dielectric oil flushes out the vaporized and condensed material while reestablishing insulation in the gap. Surface finish improves with increased frequency and reduced current. Material removal rate, surface roughness and overcut all increase with a current increase or with a frequency decrease (or longer "on" time cycles). Electrode materials frequently used are brass, copper, copper tungsten, tungsten wire and graphite. Spark gaps range from 0.0005 to 0.020 inch with closer tolerance control and slower cutting rates associated with the smaller gaps. Erosion occurs on the tool as well as the workpiece with wear ratios ranging from 0.5:1 to 100:1, depending on spark wave-shape from the power source, electrode material and workpiece material. A nearly "no wear" combination of operating parameters can be found for electrical discharge machining steel when using reverse polarity, as opposed to "standard" polarity (positive on the workpiece).

PRACTICAL APPLICATIONS

EDM cuts any electrically conductive material regardless of its hardness and is particularly adapted for machining small irregular slots or cavities. Because of the absence of physical contact, delicate structures can be cut successfully. Cutting is three-dimensional as the shaped electrode is fed into the workpiece. Because the sparks focus first on peaks and corners, burr-free cutting occurs. Multiple electrode, automatic dressing, automatic positioners and NC motion control all contribute to electrical discharge machining's versatility. Tool and die work is frequent but mass production and even transfer line applications exist. Small and/or shaped holes at shallow angles to the workpiece surface are commonplace. A recast and heat-affected layer occurs on all materials cut with EDM and needs to be removed or modified on critical or fatigue-sensitive surfaces.

MATERIAL REMOVAL RATES AND TOLERANCES

Feed rates and material removal rates range from 0.06×10^{-4} to 6.6×10^{-4} cubic inches per minute per ampere. Corner radii to 1/64 inch are common. Production tolerances to ± 0.001 inch are normal; tolerances of ± 0.0002 inch are repeatable with careful selection of cutting conditions. Finish levels are typically in the 63- to 125-microinch AA range, but deluxe equipment can attain 2 to 4 microinches AA. The recast layer can be controlled and is repeatable to a few ten thousandths of an inch. Material removal rates, surface roughness, recast layer and heat-affected zones all increase as spark intensity increases.

AVAILABILITY

EDM equipment is available in a wide range of sizes from a bench model with a few amperes capacity to 4 ft. x 6 ft. die sinkers with 3,000 amperes capacity. Automatic or NC controls are common. Automatic feed, interchangeable electrode holders and rotary turret electrode holders are available to aid electrode changing and automation of the EDM process. Multiple electrodes and multi-lead power supplies enhance the productivity of many current equipment types. Integrated systems are the usual order; thus, EDM machines can be placed almost anywhere in the normal shop. Fume vents are needed and tooling should provide for venting the gases liberated. The pulse power supply usually contains full control of "on and off" times for each discharge as well as discharge energy. Good safety practice makes it desirable to operate with the spark fully submerged in the dielectric.

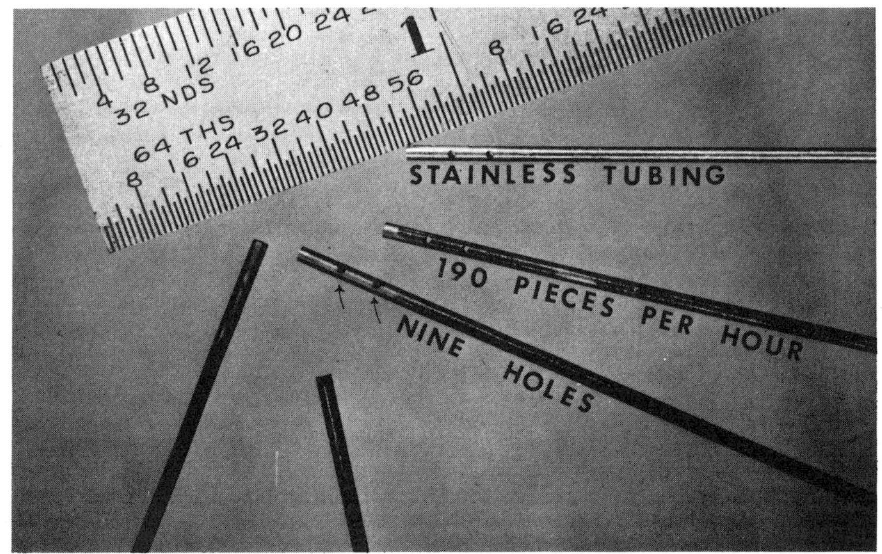

EDM of stainless steel tubing. (Courtesy of Raycon Corp.)

ELECTRICAL DISCHARGE SAWING

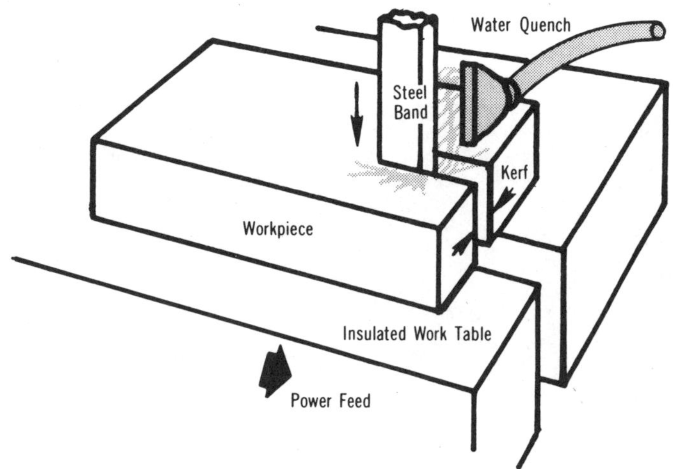

Electrical discharge sawing (EDS) is a variation of electrical discharge machining that combines the motion of the band saw with electrical erosion of the workpiece. The rapidly moving (5000 to 6000 fpm) 0.025-inch-thick, special-steel, knife-edge band is guided into the workpiece by carbide-faced inserts. A 1/32-inch kerf is formed but no controlled gap is maintained between the saw blade and the workpiece as in EDM. No dielectric is used; therefore, there is continuous arcing from the low-voltage (6 to 24V), high-current power source. Water flow quenches the arc and cools the workpiece. While the work is power fed into the cutting band, neither the band nor the work is subjected to major forces, so fixturing can be minimal. Precise adjustment of the feed rate must be made to be in exact balance with the arc erosion rate.

PRACTICAL APPLICATIONS

Fragile cellular structures can be cut from aluminum, stainless steel or titanium honeycomb. Thin-walled heat exchanger tubular assemblies can be cut. No-burr cutting produces little or no roll-over of edges on thin materials. Cuts up to 40 inches deep have been made. Only electrically conductive materials can be cut with EDS.

MATERIAL REMOVAL RATES AND TOLERANCES

Cutting rates range from 5 to 200 square inches per minute. Flatness ranges from ±0.003 inch TIR at the lower feed rates to ±0.016 inch TIR at the maximum cutting rates. The finish is an electrically etched surface; however, the arcing leaves a recast and heat-affected zone below the surface.

AVAILABILITY

EDS machines are regularly available with throats up to 48 inches deep for 26-inch workpieces.

ELECTRICAL DISCHARGE SAWING

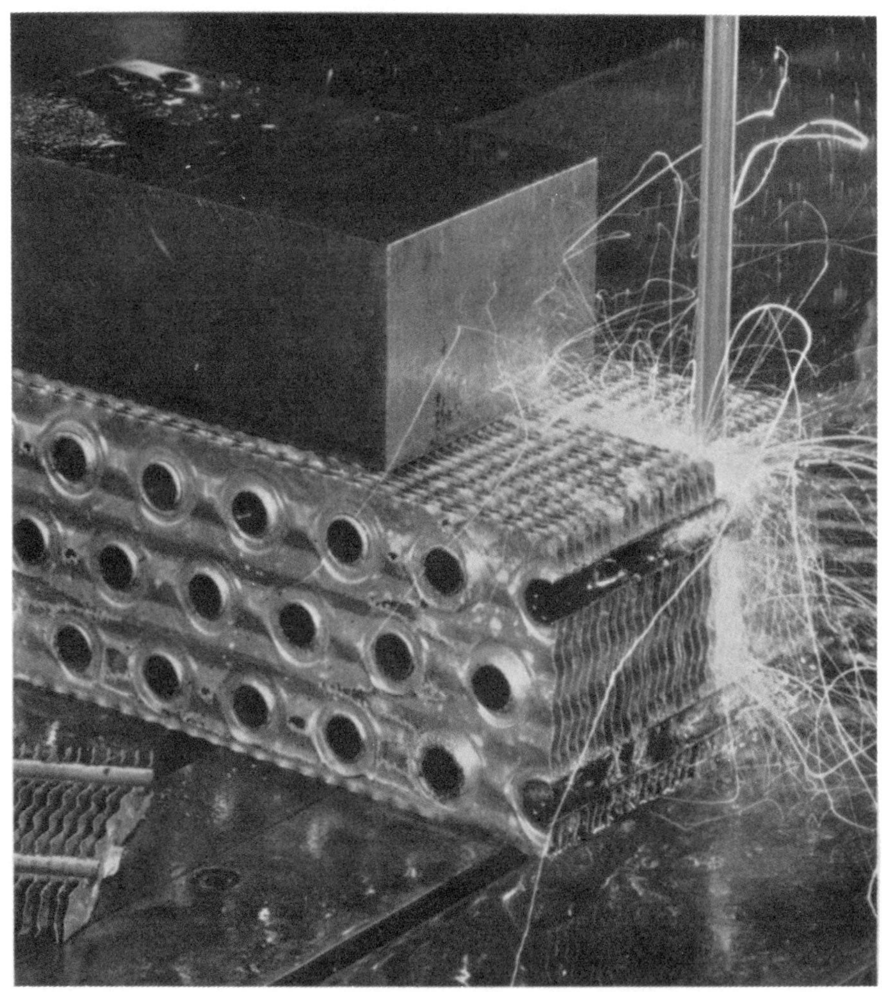

EDS of copper-tube heat exchanger at 50 square inches per minute. (Courtesy of DoAll Company)

ELECTRICAL DISCHARGE WIRE CUTTING

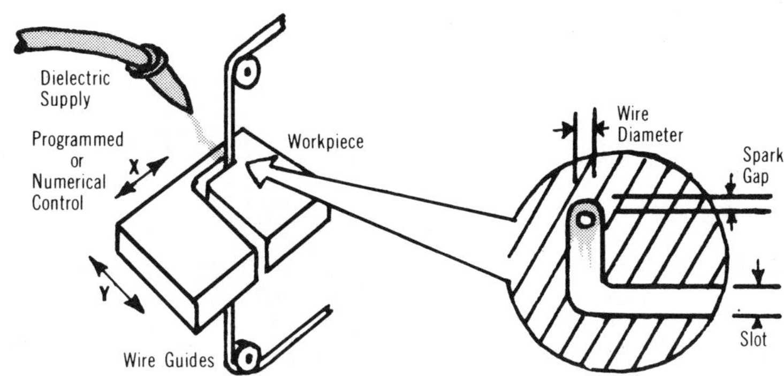

Electrical discharge wire cutting (EDWC) is a special form of electrical discharge machining wherein the electrode is a continuously moving conductive wire. It is often called traveling-wire EDM. The tensioned copper or brass wire of small diameter, 0.002 to 0.010 inch, is guided to produce a straight, narrow-kerf cut. Usually a programmed or numerically controlled motion guides the cutting while the width of kerf is maintained by the discharge controls. The dielectric is oil or deionized water carried into the gap by motion of the wire. The wire is inexpensive enough to be used only once.

PRACTICAL APPLICATIONS

The straight cut perpendicular to the major axis of the workpiece has no flaring or bell mouth and extremely tight corners can be cut with almost no radius. Punches, dies and stripper plates can be cut in any of the hardened conductive tool materials. The same NC tape can be used repeatedly for short production runs. Mirror-image profile work and internal contours from a starting hole are frequent. Stacking of sheets for multiple cutting is possible.

MATERIAL REMOVAL RATES AND TOLERANCES

Cutting of 0.001- to 3-inch-thick materials can be done at a rate of 1 square inch per hour, which on thin parts can yield cutting at 4 inches per minute. Positioning accuracy to ±0.0002 inch is normal in all metals.

AVAILABILITY

Several manufacturers regularly build EDWC equipment with NC, tracer controls and all programming accessories. Die relief, angle generators and offset controls are available on most machines along with kerf-width control via the gap setting and wire-diameter selection. Equipment is also available with cam or other mechanical programming for the wire motion as well as with standard EDM servo control for straight cutting.

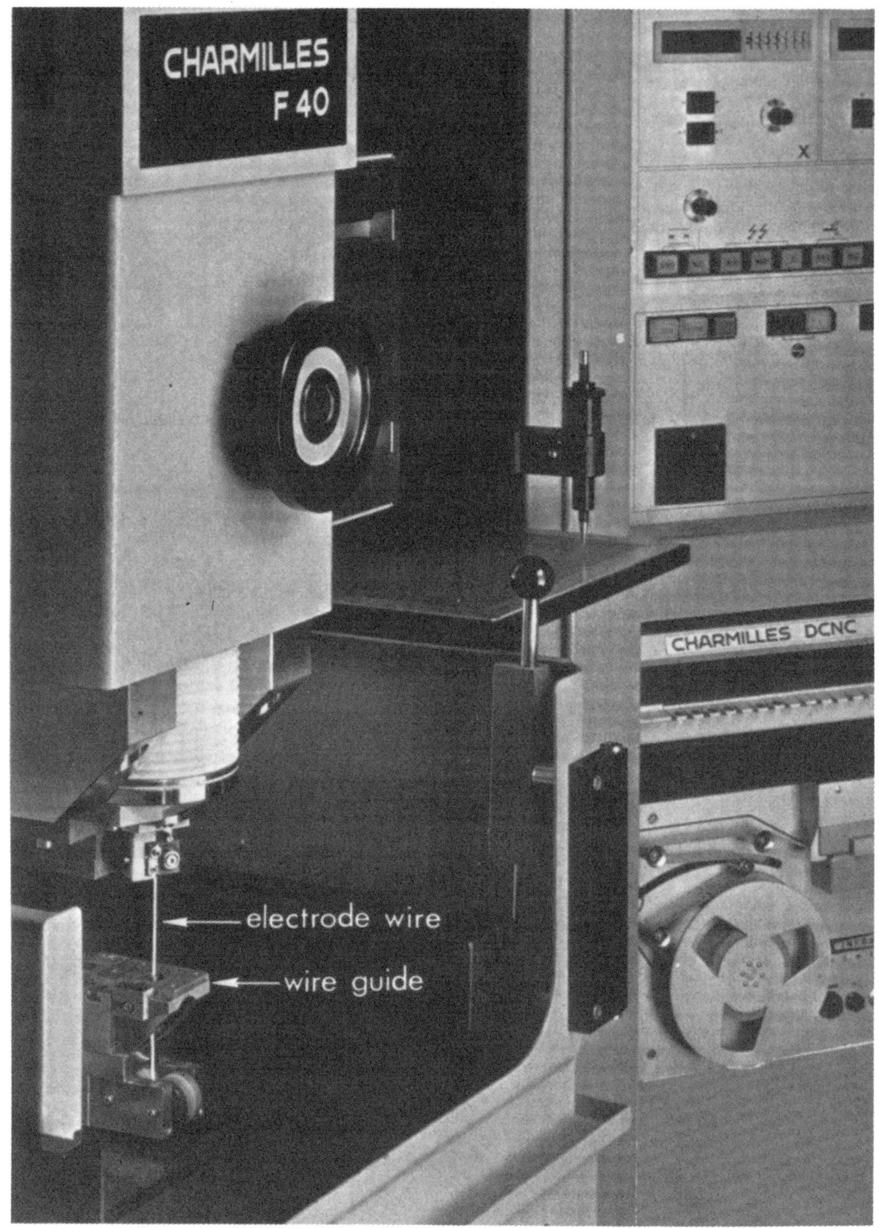

Numerically controlled EDWC machine. (Courtesy of Charmilles Corp.)

LASER BEAM MACHINING

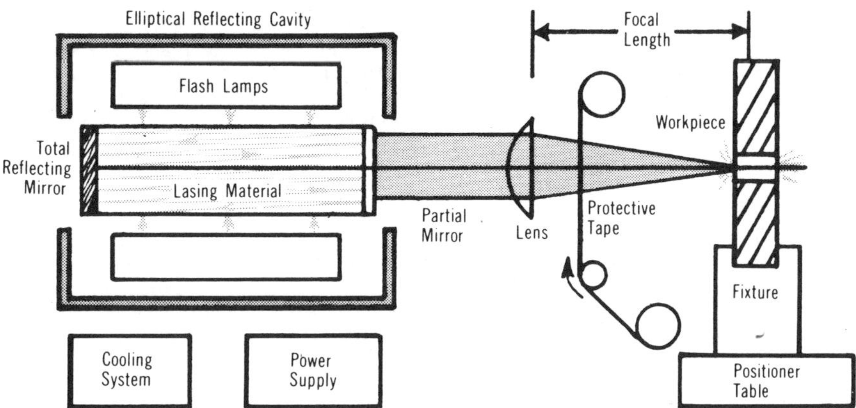

Laser beam machining (LBM) removes material by melting and vaporizing the workpiece at the point of impingement of a highly focused beam of coherent monochromatic light. Laser is an acronym for "light amplification by stimulated emission of radiation." The electromagnetic radiation operates at wavelengths from 0.3 to 300 microns. The most common wavelengths for material removal using solid state lasers are 0.69 micron for ruby, 1.06 for neodynium glass (Nd:Glass), 1.06 for neodymium yttrium aluminum garnet (Nd:YAG), and 10.6 for carbon dioxide gas laser. Of these four, only the 0.69 micron radiation of the ruby is in the visible light range. The Nd:Glass and Nd:YAG wavelengths are in the near infrared and the CO_2 wavelength is in the far infrared range. The optical characteristics of the workpiece determine which wavelength should be used.

Most of the solid state lasers operate only in a pulsed mode with greater than 10 pulses per second repetition rate common with adequate cooling systems. An exception to this is the Nd:YAG laser, which like the CO_2 laser, can be operated either pulsed or in a continuous wave to produce a continuous power output.

For pulsed operation of solid state lasers, the power supply produces a short intense pulse into the flash lamps which concentrate their light flux on the lasing material. The resulting energy from the excited atom is released at a constant frequency. The monochromatic light is amplified by successive reflections from the mirrors. The thoroughly collimated light exits through the partially reflecting mirror to the lens, which focuses it on or just below the surface of the workpiece. The small beam divergence, high peak power and single frequency provide excellent small-diameter spots of light with up to 3×10^{10} watts per square inch of power that can sublime almost any material.

During laser beam machining, the expulsed material solidifies to a dust; therefore, a cleaning system is needed in addition to protection for the lens from the molten particles. Adequate eye protection is needed for both direct and reflected laser light.

PRACTICAL APPLICATIONS

Small precision cuts or holes in thin materials can be produced by LBM. Scribing of ceramics can be done since there is no massive heat shock or mechanical contact or large forces between the tool and workpiece. LBM is not a mass material removal process; however, its operation in air at rapid repetitive rates and its ease of electrical control commend it for mass micromachining production. Multiple pulses permit hole drilling up to 50:1 depth-to-diameter ratios on 0.005-inch-diameter holes while 0.050-inch-diameter holes can be drilled through 0.100-inch-thick material. Shallow angles (15 degrees) to the surface can be drilled. Other applications include engraving, resistor trimming, sheet metal trimming and blanking. The same equipment can be used to weld, surface heat treat or machine, which makes the laser a "universal" machine tool.

MATERIAL REMOVAL RATES AND TOLERANCES

Removal rate is slow, 4×10^{-4} cubic inch per minute; however, 0.020-inch-diameter holes can be drilled in milliseconds with accuracies to ±0.001 inch in thin stock of tungsten, brass or ceramics. The hole walls will be irregular and tapered as well as having a recast structure from the heat-affected surface. Repeatability is good and yields can be improved over conventional micromachining processes. The heat-affected zones with a hardened recast layer are typical of laser machined surfaces. This recast layer can be detrimental to material properties and should be removed or modified on applications where high stresses or fatigue life is a concern.

AVAILABILITY

Several sources of laser components or systems exist. The principal equipment concern is workpiece positioning and control. Integration of an NC table with focus, beam intensity or standoff distance is common. Safety interlocked enclosures are commonly used. Bench-top equipment with a few watts capacity to computer controlled systems of several kilowatts capacity are commercially available.

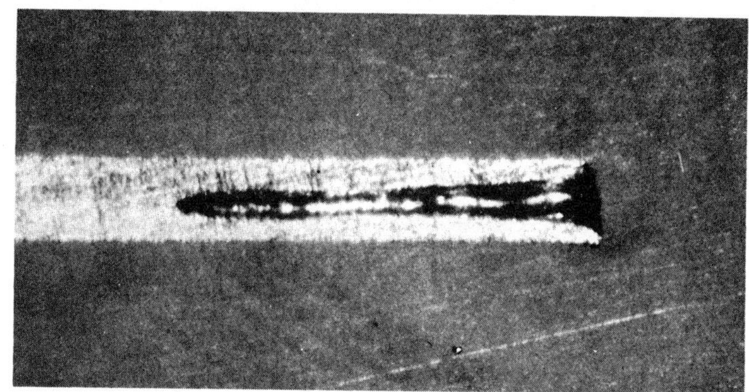

Surgical needle (0.013 inch diameter) with laser drilled 0.006-inch-diameter hole, 0.060 inch deep. (Courtesy of Holobeam Laser Inc.)

LASER BEAM TORCH

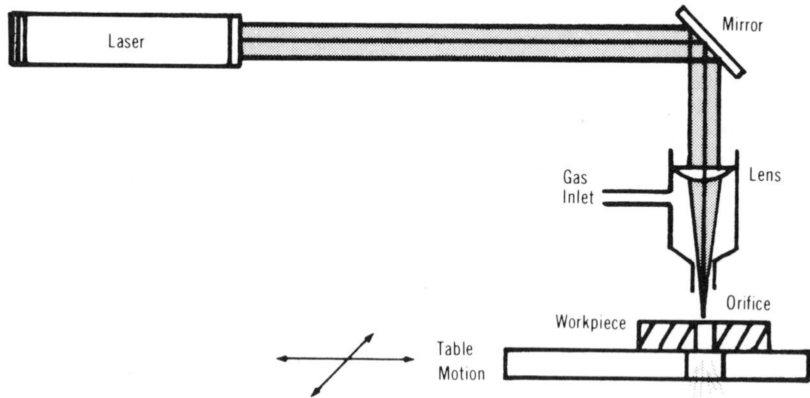

Laser beam torch (LBT) is a material removal process that utilizes the simultaneous focusing of a laser beam and a gas stream on the workpiece. A continuous-beam laser is focused on or slightly below the surface of the workpiece and the absorbed energy causes localized melting. The oxygen gas stream promotes the reaction as well as purges the molten material from the cut. Argon or nitrogen gas is used to purge the molten material and protect the workpiece when organic or ceramic materials are being cut.

PRACTICAL APPLICATIONS

The carbide nozzle and the laser's focal point must be kept at a fixed distance from the workpiece for uniform cutting. A height-sensor control is available to hold this standoff distance within ±0.005-inch tolerance. The cut is characterized by a narrow kerf and a narrow heat-affected zone. The width and quality of cut depend on cutting speed, laser power, standoff distance, nozzle diameter, type of gas and laser focal distance. Numerically controlled or mechanically guided cuts can be made in most materials. Plywood die-board slotting for steel-rule dies, titanium plates for airplanes, stacks of cloth for suits and ceramic workpieces have been cut in regular production. The low level of distortion, narrow kerf and smooth workpiece edges are advantages. Most LBT cutting is done on 0.050- to 0.100-inch-thick sheet material; however, 0.50-inch-thick carbon steel has been cut at 30 inches per minute with high-power lasers. Taper is present in most cuts. Trimming and post-cutting cleanup operations are minimal.

MATERIAL REMOVAL RATES AND TOLERANCES

The cutting rates depend on material thickness but can range from 30 inches per minute on 1/16-inch-thick tool steel to 250 inches per minute on 0.020-inch tinplate to 500 inches per minute on 0.030-inch-thick Lexan with a 1/2-kW laser. Accuracies to ±0.004 inch are obtained with heat-affected zones ranging from 0.002 to 0.010 inch wide and kerf being 0.004 to 0.030 inch wide with a 1/2-kW laser. A 1/2-kW, continuous-wave, carbon-dioxide-gas laser can cut 1/4-inch-thick aluminum at 3 inches per minute or 1/4-inch-thick titanium at 140 inches per minute.

AVAILABILITY

A wide assortment of machine shapes and controls is available with power capacities ranging from fractional kW to 15 kW. Many units employ NC for control of the cutting path.

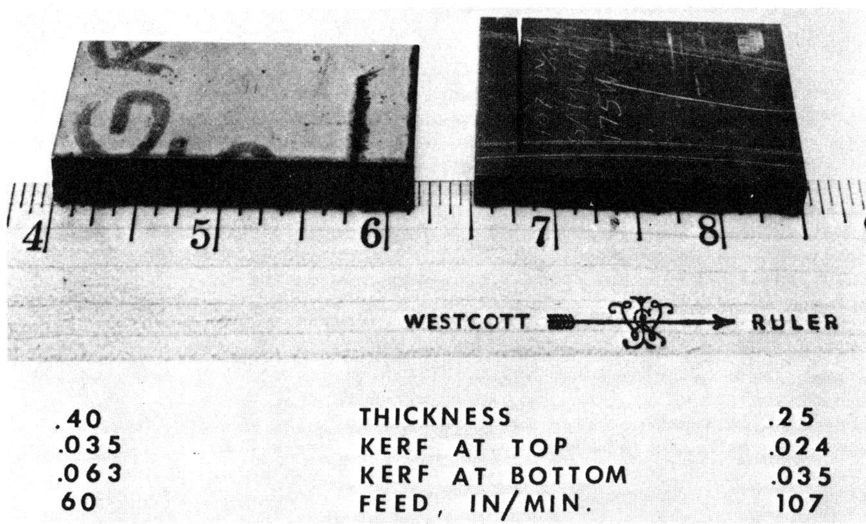

Slicing of titanium 6Al-4V with LBT. (Courtesy of The Boeing Co.)

PLASMA BEAM MACHINING

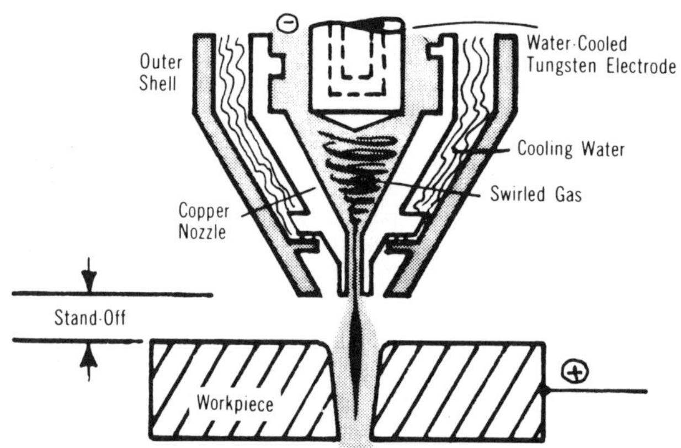

Plasma beam machining (PBM) removes material by utilizing a superheated stream of electrically ionized gas. The 20,000 to 50,000°F plasma is created inside a water-cooled nozzle by electrically ionizing a suitable gas such as nitrogen, hydrogen, argon, compressed air, or mixtures of these gases. Since the process does not rely on the heat of combustion between the gas and the workpiece material, it can be used on almost any conductive metal. The plasma — a mixture of free electrons, positively charged ions and neutral atoms — is initiated in a confined gas-filled chamber by a high-frequency spark. The high-voltage d-c power sustains the arc, which exits from the nozzle at near sonic velocity. Water injection is sometimes used to assist in confining the arc, blasting away the scale and reducing smoke. Greater nozzle life is also claimed for water-injection-type torches. Control of the nozzle standoff distance (1/4 to 5/8 inch) from the workpiece is important. One electrode size can be used to machine a wide range of materials and thicknesses by suitable adjustments to the power level, gas type, gas flow rate, traverse speed and flame angle. PBM is sometimes called plasma arc machining (PAM) or plasma arc cutting (PAC).

PRACTICAL APPLICATIONS

Profile cutting of metals, particuarly stainless steel and aluminum, has been the most prominent commercial application; however, mild steel, alloy steel, titanium, bronze and most metals can be cut cleanly and rapidly. Multiple-torch cuts on programmed or tracer controlled cutting tables on plates up to 8 inches thick in mild steel are possible. Smoothness of cut with freedom from contaminants is an advantage. Well-attached dross on the underside of the cut can be a problem. Eye shielding and noise protection are necessary for the operator and those in nearby areas.

MATERIAL REMOVAL RATES AND TOLERANCES

One-inch-thick aluminum plate can be cut at 30 inches per minute while 1/4-inch carbon steel can be cut at 160 inches per minute. The use of water injection can increase the carbon steel cutting rates to 20 fpm for 3/16-inch-thick plate. The taper on the sides of the cut ranges from 5 to 10 degrees and the kerf width is usually 3/32 to 3/8 inch. Corner radius is a minimum of 5/32 inch on thinner plates. Tolerances for slots and holes ordinarily range from ±1/32 inch on 1/4- to 1-3/8-inch-thick plates up to ±1/8 inch on 6- to 8-inch-thick plates. A heat-affected zone can range from 1/32 to 3/16 inch wide, depending on workpiece material, and depth and speed of cut. Sometimes an increase in hardness will accompany the heat-affected zone and require more time for edge-finishing operations. In many cases the cut edges may be adequate for the application.

AVAILABILITY

Programmed motion of many types from NC punched cards to optical followers is available with multiple-torch capability and cutting tables up to 44 feet x 82 feet. Cutting speeds range from 2 to 240 inches per minute.

TABLE V

SURFACE INTEGRITY EFFECTS OBSERVED IN THERMAL NONTRADITIONAL MATERIAL REMOVAL PROCESSES

Type of Effect		Typical Process Parameters	Off-Standard Conditions
Surface Roughness:	Average Range - Microinches AA Less Frequent	32 - 125 2 - 150	125 - 500 32 - 1000
Mechanical* Altered Material Zones:	Plastic Deformation (PD)	---	---
	Plastically Deformed Debris (PD^2)	---	---
	Hardness Alteration †	1.1	8.0
	Microcracks or Macrocracks	0.5	7.0
	Residual Stress §	2.0	3.0
Metallurgical* Altered Material Zones:	Recrystallization	---	---
	Intergranular Attack (IGA)	---	---
	Selective etch, pits, protuberances		0 - 1.6
	Metallurgical Transformations		
	Heat-Affected Zone (HAZ) or Recast Layer	0.6	5.0
High Cycle Fatigue: (HCF)	% change from "handbook" values at room temperature #	-17 to -96	-48 to -64

Note: A blank in the table indicates no or insufficient data.
A --- in the table indicates no occurrences or not expected.

*Maximum observed depths in thousandths of an inch, normal to the surface.
†Depth to point where hardness becomes less than ±2 points R_c (or equivalent) of bulk material hardness (hardness converted from Knoop microhardness measurements).
§Depth to point where residual stress becomes and remains less than 20 ksi or 10% of ultimate strength, whichever is greater.
#"Handbook" values from HCF testing are frequently generated from low stress ground specimens, hand or gentle machine polishing or occasionally electropolishing. These values are based upon LSG as 100% (with its minor amount of retained but enhancing compressive residual stress).

CHAPTER 2

THERMAL DEBURRING

Deburring With TEM

By A. C. Montag
President, Surftran Company.

The Society of Manufacturing Engineers has attempted to classify all deburring methods in a manner of which will allow potential users to choose the most appropriate method from a chart. Although the chart had 23 columns, it tried to answer four basic questions:

1. How does the deburring method work?
2. What can the method do?
3. What are the limitations, side effects or secondary operations?
4. What are the economics including cycle time, cost of tooling, materials and equipment?

We have polled various users of SURF/TRAN equipment and combined their responses with our own information to provide answers to these questions.

The SURF/TRAN Thermal Energy Method or TEM as it is called is probably the fastest of all the known methods of deburring. The actual time to remove even the largest burrs is generally less than 20 milliseconds. TEM works by utilizing two distinct fires. Fires always require: 1. A fuel; 2. Oxygen; and 3. Heat. If one of these three elements is missing, a fire cannot start and once started if one of these elements is removed, the fire goes out.

The Thermal Energy Method starts by placing a manufactured part with burrs into a chamber, closing the chamber and then pressurizing the chamber with a fuel gas and oxygen mixture. The first fire is started by adding heat in the form of a spark from a spark plug. This causes the combustible mixture to ignite. Unlike automobile engines where the ignition causes a piston to move, all of the energy of the fuel must be converted into heat. The heat hits everything in sight. It hits the walls of the water cooled chamber where it is carried away. It hits the main body of the part itself tending to warm it. It hits the burrs themselves regardless of their location. The heat hitting both sides of the burr soon meets itself in the middle and seeks a way to conduct itself into the main body of the part. The only way out for the heat is through the root of the burr into the part itself. Since burrs are created by the constant feed of cutting tools, the route through the root of the burr is relatively narrow - restricting the flow of heat. This restriction causes the heat to build up within the burr until it reaches its automatic ignition temperature.

After the first fire utilizing the fuel gas and oxygen goes out because all of the fuel gas is consummed, the second fire starts using the heat of the initial detonation, excess oxygen which we have placed in the chamber and for fuel, the material from which the component is made. This second fire, the burning of the burrs, will continue until one of the three essential elements for a fire is lost. This will obviously never be the fuel unless we consume the entire part and, as stated before, we have an excess of oxygen in the chamber by design. The second fire goes out only after it reaches the corner or surface of the part itself by the root of the burr. At this point, the heat is no longer restricted by the narrowness of the burr but has a suddenly wide path in which to flow. It rushes into the component itself lowering the temperature of the flame and the flame goes out.

To clear up all misunderstandings, we do not vibrate the burr off, we do not wear off the burr and surfaces of the part with an eroding action, we actually burn thin sections while merely heating thicker sections.

It is thus easy to understand that TEM can work on all combustible materials not only including ferrous and non-ferrous metals, but also thermo-plastics and rubber. You can now understand that we would have difficulty in doing a thermo-setting plastic which would not burn but simply get harder as heat is applied and you could be correct in assumming that the process requires more energy to deburr corrosion resistant metals then unalloyed metals. Understanding that the second fire goes out due to heat flowing into the component would logically mean that materials which are better heat conductors would have the fire go out prior to poor conductors. Thus, it is easier to obtain an edge break or radius on a poor conductor such as steel as opposed to brass which is a relatively good conductor.

What does this mean with regard to the size of the burr which can be removed and the edge breaking which can be performed? The TEM process has instantly removed fingernail size burrs as thick as .020" under ideal conditions. Ideal means that the remainder of the component has no sections thinner than 15 times the thickness of the largest burr and that it is capable of sustaining the detonation without damage. Usually, of course, burrs are .005" to .010" in thickness. Burr length is not a factor in TEM; it is the thickness of the burr root that is critical.

Edge breaking is a specific area of inquiry by the S.M.E. The SURF/TRAN process has put radii on steel from .002 to .060 and on aluminum, from .002 to .010. The size of the radius can only partially be influenced by the controls available to the process. The major determinants are the workpiece material and the size of the burrs. In Thermal Deburring, the radius achieved is in direct relationship to the size of the burrs. You will get larger radii where there are burrs on various edges and virtually no radius on edges where there was virtually no burr. This is the opposite of what one might expect, however, it is explainable if one considers the burrs as kindling to start a fire. The radii achieved will be consistent within ± .002" for a given material with the same size burr. In general, a sufficient radius will be put on all edges so as to prevent hand cuts from the typical ragged edges left on untreated parts. A part which is properly placed in a TEM chamber can, both outside and inside, meet that elusive specification "remove all burrs - break all sharp edges."

The equipment used to perform Thermal Deburring generally averages about 20 seconds in total cycle time for each load of parts with our factory units operating as low as 9 seconds and the slowest as high as 30 seconds. We already have trouble fitting onto the S.M.E. chart in that all other methods list a time in minutes in which a burr is abraded or eroded down to the parent metal. TEM productivity is measured by the number of components one can deburr at one time. In our average 20 seconds, although we can only deburr one large cast iron pump body, we could deburr four ordinary sized aluminum carburetors, about 30 or 40 brass elbow fittings or several hundred zinc lock cylinders. Thus rather than measuring part times in minutes, when we get above 200 parts per load we are talking about tenths of seconds per part.

The distinguishing feature of all the parts just mentioned is that the deburring problem area is internal. TEM is the fast method of removing burrs which are inside a part, possibly located in a blind hole, an inaccessible groove, or even in locations which are not readily visible. Certainly any pump, valve, metering device, fitting, etc., through which air, water, gas, oil or any liquid must flow is an ideal candidate for the TEM process. TEM is an excellent way of eliminating the loose debris often found internally in such components. TEM, for example, is now used by virtually every major carburetor manufacturer in the world as the method of providing consistent quality to meet the new environmental standards. It is gaining in popularity among hydraulic and pneumatic component manufacturers.

The limitations to the process generally result from the affects of the shock and heat waves connected with the detonations to fragile or thin parts. There is a stress relieving characteristic equivalent to dropping the parts from a height of over one foot which is especially noticeable on highly stressed parts such as formed or stamped sheet metal parts. Air hardening steels which have thin sections can be troublesome if it is impossible to shield or heat sink the thin sections and if the part is not to be heat treated later. On rare occasions, customers will not accept the recast layer under a removed burr. Since the burr was burned off in a gaseous state, there is a portion of the burr which was liquid which resolidified as the second fire went out. Our tests have shown that this recast layer is generally thin enough so as not to affect tool life on subsequent machining, if any. The process is also not recommended for materials which have been previously hardened in that the surfaces occasionally develop heat check cracks.

Original S.M.E. papers list it as a side effect the fact that we convert burrs into oxides since this required, in many instances, the secondary cleaning operation. In actuality, all of the tumbling and vibrating methods must also clean the parts to assure that there are no bits of the burrs remaining in them and the chemical methods must wash the parts to remove electrolyte or acid. We have gained knowledge of a line of liquid solvent which will disolve the oxides which the TEM process forms without reducing the size of the main surfaces. This oxide removal can be easily performed up to three days after the thermal deburring because the oxides which we form are not yet married to the part. In other words, we distributed oxides from the converted burr all over the part as opposed to forming oxides all over the surface. If you anodize your aluminum, dichromate your zinc, brite dip your brass or harden your steel, you will need no additional cleaning equipment along with thermal deburring since these processes remove the TEM oxides.

The process can and is used for the removal of flash, however, it will not remove parting line flash entirely since the second fire goes out as the flash widens at its root. The results are generally acceptable for functional purposes, however, if the beauty of an abrading or buffing action is required, TEM will not fill the bill. On the other hand, you have the assurance that TEM will not abrade large flat surfaces of a part while removing most of the parting line flash.

To discuss the economics of TEM, it is necessary to visualize the equipment used. A typical unit using natural gas is shown in the schematic, Figure 1. Basically, the unit is a toggle type press having a C-frame capable of a squeezing force of 250 tons. This force is used to hold the lid on the pressurized chamber. The unit happens to have the chamber upside down and utilizes six lids which are raised up to the chamber. These lids which we call lower closures are the items upon which parts are loaded. Depending on size and amount of automation, units such as these can cost between $60,000 and $250,000.

Customers with large rectangular parts with internal burrs utilize no tooling at all and they simply pay for the fuel costs which, including rental of the tanks from a supplier, average between $.01 and $.015 per detonation.

Customers with many small parts do have to contain them in some sort of basket to prevent them from spilling over the sides and these stainless steel baskets generally cost below $100 each. Customers with parts which have external burrs which are to be removed simultaneously with the internal burrs must raise the parts up in a manner so that the external burrs are "in the gas." This requires some form of pin, post or nest which, although it must be made of stainless steel, does not require the precision of an ECM electrode, the form fitting of the abrasive putty machines or the complexity of the holders on spindle finishing machines. Some customers for a few thousand dollars, have made standard subplates and pin sets capable of handling several parts of varying sizes and shapes.

Just as all the deburring methods utilizing an abrasive wear out various components, various seals for TEM are subjected to wear by the process and must be regularly changed. We like to compare our chamber and mixing valve seals to cutting tools on a lathe or milling machine when stressing the need for regular, scheduled changes. It is our hope that rather than having this changing time classified as downtime, it will be classified as set-up time as it is on normal machine tools.

It is difficult to give an average value for the cost of seals, spark plugs, thermocouple tips, etc., for the process because their use is directly proportional to the energy level which we are operating. For example, a zinc manufacturer would use relatively low energy levels in which a set of such components could last well over 30,000 detonations which, in turn depending on size, could mean 120,000 to 240,000 carburetors. This could be about two cents per detonation or mills per part. On the other hand, a customer deburring a stainless steel part one at a time in a special fixture at a very high energyllevel could require new seals before 5,000 detonations requiring costs in excess of $.05 per workpiece.

These figures are meaningless without a comparison with other methods. In general, if a customer has an automated method which guarantees him the quality necessary, it will generally be more economic to stay with it rather than switch to TEM. If, however, quality demands that manual deburring be utilized and if the customer is using four or five people for this purpose on a fulltime basis, the thermal method will probably be able to show significant savings. The greatest savings of course will be in the lack of warranty charges for products that went wrong because of burrs that you could not reach at all.

One of the nicest features of TEM is the ability to quickly find out if it's suitable for a certain component or not. Anyone having a component which will fit into the 10" diameter by 6" high chamber can simply send them to the factory for no charge trial. There is no special media which needs to be developed or electrodes which must be manufactured.

The results on one size component will be fairly typical of other parts in the same general family.

We believe that if components are designed correctly with TEM in mind that TEM can satisfactorily meet all realistic deburring requirements for internal and external burrs. We feel, in fact, that TEM will offer a design freedom for both product design and processing never yet achieved. It will allow you to make use of the fact that gas goes everywhere and, therefore, you can be assurred that wherever there are burrs TEM will find them and eliminate them.

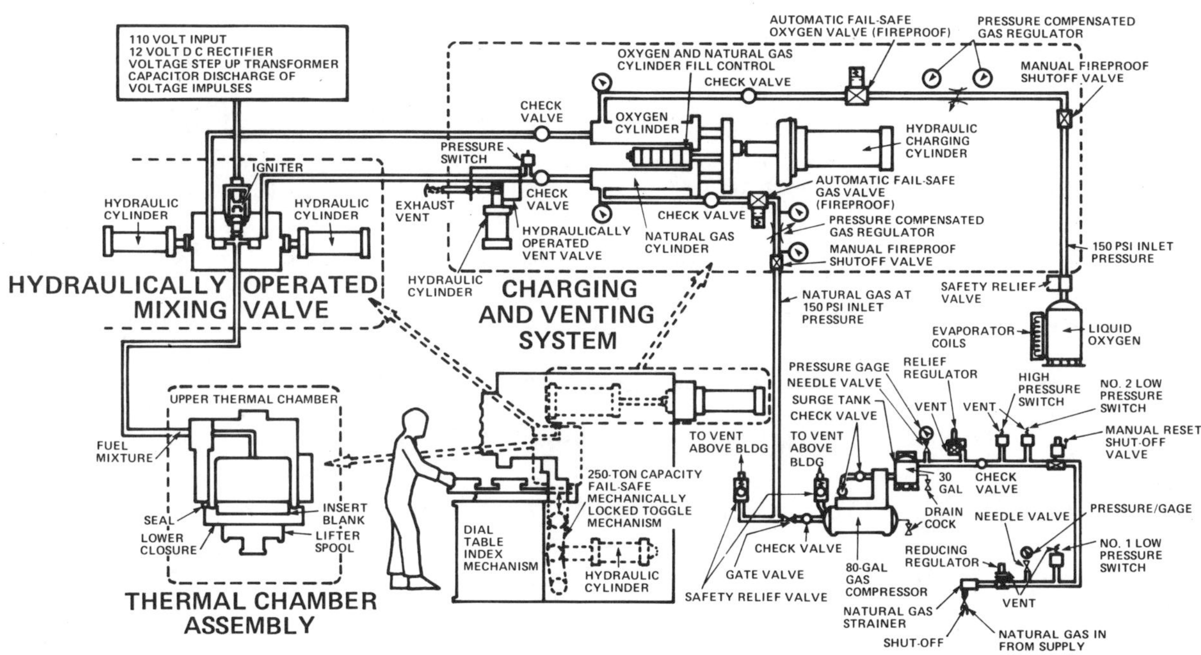

Figure 1

Assuring Quality With The Thermal Energy Method

By A. C. Montag
President, Surftran Company

The Thermal Energy Method is still considered a non-traditional way of removing thin sections from workpieces. It is so because of its newness. There are still only about 130 units in operation. Unfortunately, amongst any group of engineers a good percentage will still not know how the method works.

First of all, let us establish why today's quality demands require that thin sections, burrs, flash and particulate matter be removed from components. A half dozen reasons come to mind almost immediately. First, the obvious desire to avoid loose pieces from causing mechanical damage, such as chips falling between rotating gears. Secondly, one must obviously remove a raised edge if it restricts mechanical motion of a valve spool moving within a valve. The third would be a liquid or gaseous restriction by particulate matter in filters, orifices, etc. Fourth would be burrs, flash or fins causing turbulence or non-laminar flow in a liquid or gaseous system. Fifth, avoidence of sharp edges which cut the hands and fingers of persons handling the components; and last, the appearance of the component - which sometimes make it marketable or not. The non-traditional machine processes being discussed at this conference all have their special reasons for existence. Most of them, however, have the common advantage that they do not create at the outset the six problems stated above which are traditional and we think unavoidable with conventional cutting tool type machining. TEM, if properly used, can be the means to avoid non-traditional machining in some cases while achieving the same kind of results.

So what is TEM? It consists of placing manufactured parts containing undesirable thin sections into a chamber, closing the chamber and then pressurizing the chamber with a fuel gas and oxygen mixture. Having two of the three elements necessary to maintain any fire, we now need only a third -- heat. In TEM the initial heat is supplied in the form of a spark from an ignition system. The spark causes the combustible mixture to ignite and all the fuel gas is consummed in approximately two milliseconds to form a wave of 6,000°F heat. The heat hits everything within the chamber. It hits the water cooled walls of the chamber where it is carried away; it hits the main body of the part itself tending to warm it; it hits the burrs wherever they are -- within blind holes -- obvious external edges -- in intersections of small intersecting holes deep within the part -- and even in positions which cannot be reached by hand or seen by eye.

For TEM, the main difference between the burr and the main body of the part is the surface area to mass relationship of the two. Whereas heat hitting the main body of the part slowly warms it because it is flowing toward the middle of the part, heat hitting both sides of the burr meets itself in the middle while attempting to flow into the main part through the root of the burr. Due to the thinness of a burr produced by a cutting tool utilizing constant feed, the heat cannot flow into the main part rapidly enough to prevent the burr from reaching its automatic ignition temperature. Thus, the heat created by the initial ignition of the fuel gas is used now to start a second fire combining oxygen with a new fuel -- the burr material itself -- be it steel, brass, plastic, whatever.

Once started, the burrs will continue to burn until the fire loses either fuel, oxygen or heat. We have assurred the abundance of oxygen in our initial fuel mixture and obviously we will never run out of fuel since the component now comprises the fuel. The burr generally goes out when the corner of the main part is reached at which point the heat within the burr can easily flow back into the main part.

That's it! Its like burning the wick of a candle. There are, however, some limitations -- first of all, your components must be able to fit into the size of chamber we have available. This would be 10" in diameter on current models. Secondly, we can eliminate machine burrs quite well but not triangular thin sections such as flash or the burrs quite commonly found on stampings. In fact, on flash we simply round the thinner edges. Because the process has no knowledge of what is good or bad, we need a range in which to work usually requiring that the items you wish to remove be 10 times thinner than the sections you wish to keep. The process demands that the thin sections to be removed be "in the gas," therefore, they cannot be touching one another or in holes filled with cutting oil or coolant. Lastly, the process is not for light or fragile parts due to the movements caused by the shock wave used to produce the initial heat. For such light, fragile parts we recommend the other inside deburring method that takes off only burrs, ECM. In fact, SURF/TRAN represents Robert Bosch GmbH of Germany, the world's largest producer of ECM equipment.

The equipment used for the TEM process consists of a method of getting the gases in the chamber and holding the chamber closed during detonation. A high production type which averages about 22 seconds for floor-to-floor time for a basket full of parts is shown in Figure 1. A slower one minute cycle shuttle type for larger components is shown in Figure 2.

Now let's see how we can utilize the capabilities and limitations we have just mentioned to make true progress in improving quality. Our first example would be in the diecasting of zinc and aluminum components. Currently, such parts, after coming from the diecast machine, are put in a trim press which attempts to remove sprues, risers and flash. If the designers know that the part will be thermally treated after machining, the diecasting can be so designed that the thick sprues and risers are put in a plane capable of being reached by the trim press and parting line flash which, if rounded, does not detract from marketability is left on to be removed by the thermal process. The quality improvement is that all residual flash remaining is glued to the component "for keeps." There's no possibility of loose bits of partially trimmed flash getting into the mechanism or cutting anybody's hands.

Our next example - A lock cylinder. Typical "before" components show the need of deflashing and some internal machining. TEM cleans out the unwanted flash and allows perfect functioning. This example illustrates good design concerning the "10 to 1" rule for sections which must be eliminated. A poor design which we could not treat properly usually has massive items connected to the main body of very thin sections. Usually the process would attack the thin section and the entire piece would be disjoined from the main component. This illustrates how designers can be helpful if they think through the entire process from the casting or raw material stages through machining and also, hopefully, finishing.

Our next example is a classic in highlighting the changes of thinking required when new processes are developed - A typical valve body containing a bore in which a spool must operate. Into the bore a smaller hole is drilled to carry a fluid or gas. Today's engineer or process man would state that the proper method of producing this part would be to first complete the larger diameter bore and second to drill the much smaller diameter hole into it. Why does this hold true? Every process man knows that he wants the burrs created from the small drilled hole to end up in the large bore where he has a chance of removing them by entering the large bore again with some sort of knife or abrasive tool. The fact that the work of the hand tool can be ununiform causing turbulence or lack of a sharp cut-off for the spool is ignored because of the hopelessness of the situation if the large bore gets down below a half inch. A designer who plans on using TEM, on the other hand, can drill the small hole into the solid (which incidentally is better for the small drill anyway) and can then bore the large hole last. This pushes the burr of the large hole into the small hole but the TEM process can round it off in such a fashion that laminar flow is assurred and that spool movement is unrestricted.

This little trick of processing allows us to claim that we can do deburring where we do not restrict mechanical motion. In fact, with this change of processing we feel that we can meet all six of the common thin section removing reasons -- in other words, we can remove all burrs and break all sharp corners.

Next is a component which we regularly free of "particulate matter." It is a portion of an intake manifold system which could conceivably allow metal particles to enter into the very precision machined interworkings of the engines produced by a major U.S. company. These components, since they have no internal passages (the usual reason for TEM), are generally treated by conventional blasting, vibrating, tumbling and what have you. All of these methods tend to hammer-in small bits of component into the skin in unpredictable positions. In operation, these small particles can fall out and TEM is used to shock them out of position and turn them into the oxide of their original metal. In other words, on this particular component we turn the bit of metal into aluminum oxide. Our second step as in all TEM operations requiring cleanliness is to dissolve the oxide of the metal in a specific solvent which dissolves the oxide without attacking the metal. After this wash plus an ultrasonic rinse, the parts are dried and placed in dustproof plastic jackets where they are kept until assembly.

The feature which we think will eventually make TEM a traditional method is its universality and ease of try-out. The gases surround every component regardless of its shape, type, part number or stage of machining. Generally speaking, more than half of the components can be processed without using any type of holding fixtures. Our "media" never changes. It's always natural gas or hydrogen.

To try-out some parts, we put them in the chamber, choose a low pressure and fire away. If we have not removed the largest, undesirable thin section, we increase the pressure and try again and again until we have removed the largest unwanted sections, burr, flash, etc. If at that stage we have not damaged the part by eliminating a desirable portion or if we have not distorted the part, the trial is successful and can be repeated consistently for millions of cycles. If we have damaged something on the part, we might have to take off the largest thin section by hand and lower our pressure and only go after the smaller burrs, etc., which, in fact, are usually the more difficult to remove or find.

TEM is available to everyone by purchasing the equipment or on a contract basis in a few major industrial centers.

FIGURE 1

FIGURE 2

CHAPTER 3

PLASMA

Reprinted from: Machine and Tool Blue Book, November 1975

Plasma-arc cutting and punching machine improves productivity for fabricators

By **Herman Reichardt**, Contributing Editor

A piece part in the machine with squares and circles cut out by the plasma-arc under NC control. Plasma-arc unit is in the large circular housing to the left of the punching head.

The marriage of two methods, cutting and punching of sheets and plates, represents an important cost reduction and productivity improvement union, and when NC is added you have the first numerically controlled hydraulic punching machine with a plasma-arc cutting attachment. There are some worthwhile benefits to be derived from this system:

1. Material handling costs are reduced by 50 percent because piece parts are loaded and unloaded only once in the machine for cutting and punching rather than twice when two separate systems are used.

2. Both are controlled by one NC tape.

3. Plasma-arc cutting is accomplished at a rate of 200 ipm; up to 1/2-inch thick high-strength material can be cut.

At the 200-ipm plasma-arc cutting rate the only part of the machine which moves is the table, and then it is moving at only one-third the rated traverse speed used when punching holes. Another important feature of the plasma-arc is that it is the only fabricating tool which works faster in stainless steel and other hard materials than it does in mild steel. The high speed of cutting does not allow heat to penetrate more than a few thousandths of an inch in from the edges of the cut. This produces little or no distortion in the sheet or plate.

Plasma-arc cutting is done with a high-velocity jet of high-temperature ionized gas. The relatively narrow plasma jet melts and displaces the workpiece material in its path. Because plasma-arc cutting does not depend on a chemical reaction between the gas and the work metal, and because temperatures are extremely high, it can be used on almost any material that conducts electricity, including those that are resistant to oxy-fuel gas cutting.

The plasma jet heats the workpiece by bombardment with electrons (anode spot effect) and by transfer of energy from the high-temperature, high-energy gas. This is accomplished by ionizing a column of gas with an electric arc and forcing both the gas and the arc through a small orifice. The resulting plasma-arc produces temperatures up to 30,000 degrees F. When this high temperature, high-speed plasma stream and electric arc strike the workpiece, the heat rapidly

Panel in the upper part of the photo has been both conventionally punched and cut by plasma-arc. The cut-out portions in the foreground can be used for other purposes.

melts the metal and the high velocity gas blows away the molten metal. This process makes a clean, high-speed cut with little or no formation of slag, requires no preheat and produces a minimum heat-affected zone. Plasma cuts often require no further finishing. The first prototype of this machine, made by W.A. Whitney Corp., was shown at the International Machine Tool Show, 1974, and is now in full production.

A large number of parts can be made from one sheet even though the piece part is too large to be blanked by the punching process. Shearing operations can either be reduced or even eliminated.

The punching capacity of this 40-ton machine remains high with a maximum hole capacity of 5 inches. The hydraulic system produces full tonnage throughout the stroke of the ram. Table speeds of the machine have been increased from 400 ipm up to 600 ipm which, in turn, also increases the hit rate.

Maximum cutting flexibility is achieved by using a contouring control which has a programmable feed rate for changing the cutting speed on different thicknesses. Linear and circular interpolation adds to the ease of programming and reduces the length of tapes required for the average program. In anticipation of changing over to metric in the future, an inch/metric feature is offered as a standard. ●●●

Reprinted from: Production, May 1977

Plasma-Arc Cutoff Machines Chop Labor 60 Percent

Three operations . . . sawing, milling, and deburring . . . are eliminated as custom-built machines contour-cut ends of tubing for crane booms. Cut ends are ready for welding

Tube cutoff operations are considered rather routine in many plants, but not at American Hoist & Derrick Co., Industrial Brownhoist Div., a major manufacturer of locomotive cranes, crawlers, and other heavy lifting equipment. Industrial Brownhoist has a special tube cutoff requirement; both ends of every 2- to 6-in. diameter tube section used for boom "lacing," over 14,000 each year, must be "saddle-cut" to allow the tube ends to wrap around other tube bodies at assembly.

Two years ago, a three-man tube cutoff operation was replaced when division engineers, at the Bay City, MI, plant, designed and built a semiautomatic plasma-arc tube cutoff machine, capable of cutting the special saddle configuration on 2- to 4-in. diameter tube sections. The machine cut both ends of a lacing simultaneously and required only one operator. So attractive were the labor savings that, six months later, a second machine was built to cut 3- to 6-in. diameter tube sections. Today, the

Two custom built plasma-arc machines contour-cut both ends of tube sections in 4 to 15 seconds. The machines cost about $20,000 each to design and build

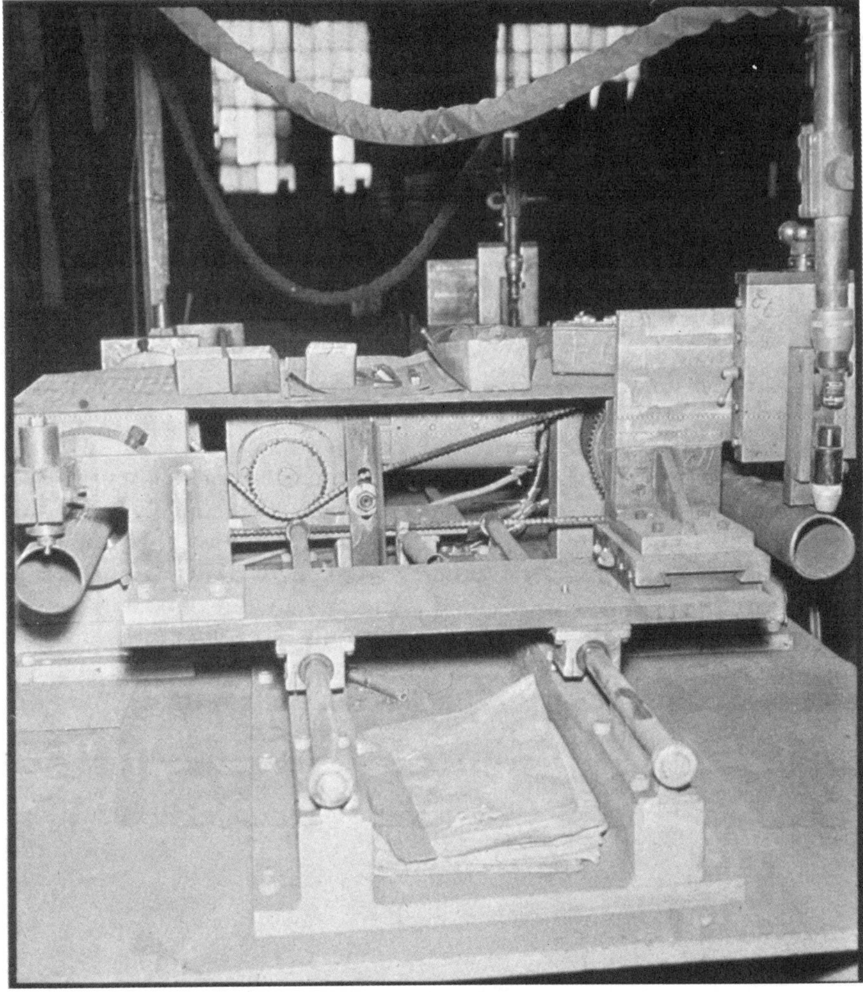

Cutoff machine tooling is designed for easy changeover; the operator can change the master tube section and adjust the carriage and stylus to the new length or tube diameter in a matter of minutes

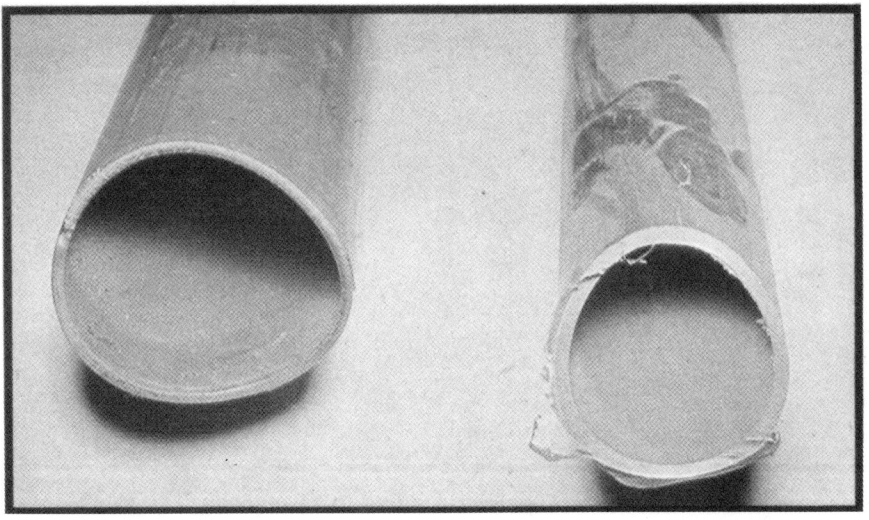

The unusual tube-end contours require no deburring or deslagging when cut with the plasma-arc cutoff machines. The old process left burrs which had to be removed before welding

two machines cut over 90 percent of the annual tubular "lacing" requirement.

The so-called "lacings" are the cross members in a boom assembly which support the major structural members. Industrial Brownhoist has over 50 different tubular lacing section designs that range in size from 3- to 13-ft in length and from 2- to 6-in. in diameter.

". . . we were not aware of any standard machines like them, so we did the development work ourselves."

Prior to the installation of the two plasma-arc tube cutoff machines, the special saddle-cut was completed in three operations: friction saw cutoff, mill, and deburr. Edward Ralph, chief manufacturing engineer, explains: "The old process took so long and the limitations were such, that we couldn't keep up with production schedules. The cutoff, mill, and deburr operation labor costs were $1.80 per lacing. In addition, we had material costs for the cutoff wheels and milling cutters. There were three men on that job, and with so many different lacing designs, we were constantly changing over; a process that usually took hours."

Douglas Hillman, chief tool designer and originator of the plasma-arc tube cutoff concept at Industrial Brownhoist, recalls: "When the two plasma-arc machines were designed and built, we were not aware of any standard machines like them, so we did the development work ourselves."

Both plasma-arc cutoff machines are equipped with a pair of template-controlled torches. However, the machines differ with respect to template design: The first machine, capable of cutting 2- to 4-in. diameter tubing, uses a variety of single purpose templates. The templates are actually lacings that have been machined to part print specifications of length and tube end contour; they could be described as "master lacings." The second machine, cutting 3- to 6-in. diameter tubing, uses one adjustable template instead of many single purpose masters.

In operation, the machines rotate the template and workpiece about their longitudinal axes. At each end of the template, a carriage holds a stylus and torch. As the template and workpiece rotate, the stylus follows the template tube-end contour and imparts the same motion through the carriage to the torch. In this way, the rotating workpiece is cut to the exact length and tube-end contour as the template.

Tube wall thicknesses of up to 2 in. can be cut with the larger machine; the smaller machine can cut up to ½ in. Both machines are equipped with plasma-arc power supplies and torches supplied by Thermal Dynamics Corp. The smaller machine, for example, has two PCH4 torches and both PAK 20 and PAK 22 power supplies, 100 amp models.

According to Gary Kester, plant engineering technician, the plasma-arc cutoff machines use "one set of consumables every full shift." For each torch, the "consumables" consist of an insulator, nozzle, and gas diffuser; the total cost each shift is about $29 per machine.

The plasma-arc tube cutoff machines have eliminated tube end deburring

Labor Savings. "Tubular lacing cutoff labor costs have dropped about 60 percent since we installed the two plasma-arc cutoff machines. Each tubular lacing cut the old way cost us $1.80 for straight labor; it costs only about $.65 per lacing with the plasma-arc machines. One man operating a plasma-arc machine can produce over 100 lacings per shift, even with changeover. Three men using the old process can't match that.

"However, about 10 percent of our tubular lacing requirement cannot be saddle-cut on the plasma-arc machines; the saddle-cut contour is so extreme on these few designs that the machine's two spring-loaded carriages cannot react fast enough to produce an accurate cut. Therefore, we still cut some lacings with the old process," reports Hillman.

"A significant portion of the labor savings has been generated by the elimination of the deburring operation," adds Hillman. "In most cases, no manual deburring or deslagging is required when lacings are cut on the plasma-arc machines; the parts are ready for welding."

No Slag Removal. Each plasma-arc torch contains a gas diffuser that directs inert gases at the cut. Gases such as nitrogen and/or CO_2 are used to "blow-out" the cut, provide cooling to the torch nozzle, and direct the flow of slag. Standard gas diffusers were installed on the machines when they were built. However, when the machines were put into production, it was discovered that one end of every lacing was free of slag but the other end always required slag removal.

Industrial Brownhoist and Thermal Dynamics engineers developed a special gas diffuser to eliminate the problem. The special diffuser "swirls" the gases in the opposite direction as compared to a standard diffuser. They substituted the special diffuser for the one standard diffuser on each machine that was "throwing" the slag onto the workpiece. Today, both machines are capable of cutting virtually slag-free lacings since both torches throw the slag toward the scrap end-piece instead of the lacing.

Hidden costs have always been a concern of management, but sometimes an economic venture may hold elusive hidden benefits as well. The labor and material costs for both the old and new processes are quantifiable, as the records show. But the new process also provides consistently accurate tube-end contours that probably save welding time and increase the strength of the boom assemblies—but these things are difficult to measure.

One thing is clear; Industrial Brownhoist has recognized the relationship between a manufacturing organization's profitability and its inventiveness. □

Thomas J. Drozda

Plasma Arc Cutting

By John E. Barger
Chief Manufacturing Engineer
Grove Manufacturing Company

The term plasma, as used in physics, means a stream of ionized particles. This may be likened to a streak of lightening which ionizes the gases of the atmosphere and heats them to incandescence. The electric arc is capable of ionizing both solids and gases. Substances are usually thought of as existing in three states; either as liquids, solids or gases. Ionized substances are sometimes thought of as being in a fourth state of matter.

The plasma torch provides an electric arc between a tungsten electrode and a water cooled copper nozzle. Gases such as nitrogen, hydrogen or helium are forced through the arc and nozzle with the result that they are heated and become ionized and the stream from the nozzle is therefore a plasma stream of ionized particles.

While the plasma torch may be used for both welding and cutting, the text of this paper will be directed toward the cutting properties as related primarily to ferrous metals.

The technical break through that permitted the cutting of square corners in carbon and low alloy steel occurred several years ago. This was achieved by combining the plasma cutting process with an N/C director.

This, in effect, opened the door for industry to apply a new machining concept that minimizes many traditional methods such as sawing, drilling, milling and shearing.

PLASMA MACHINING CARBON T-1 AND VAN-80 STEELS

The problem of consistently cutting shapes varying in size from 1/8" to 3" x 4" (3.175 mm x 7.62 cm x 10.16 cm) to 3/4" x 42" x 28' (19.05 mm x 1.07 m x 8.53 m) and maintaining an overall tolerance of +/- 1/16" (1.59 mm) and an edge finish of 250 RMS was solved by installing a numerically controlled plasma machine.

Machine make up and support systems consisted of:

A. Basic cutting bridge

B. N/C director

C. Plasma torches (Figures 1 & 2)

D. Power supplies

E. Water coolers

F. Festoon system

G. Water tables

H. Material handling system

TAPE PREPARATION FOR PLASMA CUTTING AND RELATED PROBLEMS

In N/C flamecutting, there are several programming problems that are unique to the programming of plasma arc. Most of these problems are caused directly or indirectly by the speed with which plasma cuts or by the bevel generated on the piece part by the plasma arc.

One major problem is the rounding of outside corners and the burning off of sharp end points on plasma cut parts. This is caused by the "following error" of the N/C machine. Following error occurs because the flame cannot keep up with the programmed path of the torch and when a direction change takes place, the flame tends to round the corner. The faster the machine is going, the more pronounced the rounding of corners becomes. In the case of a part with a sharp end point, this can sometimes cause the part to be as much as 1/8" (3.175mm) to ¼" (6.35 mm) short of the actual programmed dimension. Following error can be corrected by using programmable pause or dwell codes. With proper use of such codes, it is possible to cut relatively sharp and distinct outside corners.

The cutting of inside peripheral corners and of inside cutouts is another programming problem, also related to the following error. A perfect inside 90° corner is virtually impossible to attain. On one hand, if no pause is programmed in the corner, a rounded or radius effect, sometimes very pronounced, will be produced. On the other hand, if pause is programmed in the corner, often a blow hole or gouge is burned into the part. This not only detracts from appearance but also could cause a point for weakening that could lead to part failures.

Experimentation with the individual machine is necessary to determine which method produces the most satisfactory results.

The location of a pierce point and burn-in and burn-out is often the most difficult step of programming any plasma cut part. It is necessary to choose a pierce point for a burn-in to get the most economical use of the raw material. The job is further complicated by the fact that the point that may be the most economical is often times not the most feasible from the stand point of getting a good plasma cut part. For example, burning in on a radius is a bad practice especially on thinner material because oftentimes the part drops out before the cut is complete and a tip is left. The bevel caused by the plasma arc makes it good practice to try and make the starting and finishing cuts as close to being perpendicular to each other as possible. Semicircle burn-ins can be used on radii and angled lines but they sometimes let slight gouges in the part.

N/C flamecut machines with kerf compensation capabilities greatly simplify part programming but they also create other problems. Kerf compensation units make it possible for the programmer to program the part exactly as it is shown on the engineering print. There is no need to take into consideration the width of the flame which often varies due to raw material or the burning tip. Kerf compensation is an additional factor which helps multiply the effort of following error on inside corners and cutouts. Kerf compensation is also the major reason for having a burn-out.

Once a part is cut, if the kerf is shut off before the torches are turned off, the flame will cut back into the part and make a nick. The burn-out gives the flame a place to cut into without hitting the part.

The bevel of the plasma arc also makes it impossible to burn certain parts by the plasma method. In such cases, it is necessary to process these parts across mapp gas machines.

N/C tape preparation for these machines is done mainly by computer assist with a minimum amount of manual programming. Our system consists of a mini-computer, 10 CPS terminal, 75 CPS high speed tape punch and reader, high speed line printer and a graphic plotter. The system and the necessary software are leased from an N/C computer assist company. A time sharing backup system is also available.

To program a detail part, a text of geometric definitions defining the part from the engineering print and a series of cutting statements are fed into the computer. The computer processes the input and aids in the debugging of the program. The graphic plotter is used to further debug and verify the program. It is also a valuable aid in determining nesting patterns and economical raw material utilization. After the debugging is complete, machine tapes are pulled and sent to the shop. This process takes anywhere from ½ hour to 3 or 4 hours, depending upon the complexity of the part. Average length of tape is 14' (4.2 m).

LEARNING CURVE

During initial startup of this installation, we were confronted with many problems over and above what we considered normal for conventional machinery.

Problem areas to be considered are:

- Controlled amperage for plate thickness
- Controlled standoff from plate material
- Simultaneous torch firing
- Cutting speed
- Initial penetration point to minimize plate bowing
- Kerf compensation
- Tip size
- Gas flow
- Water Flow

As we went about solving these problems, a series of charts were developed. Continual fine tuning of the system provided us with information illustrated in chart on page 11.

QUALITY OF FINISHED PARTS

An experienced operator will consistently produce completed parts with an edge finish of 200-300 RMS with minimal dross adherence to underside of material. (Figure 3) NOTE: Plasma cut shapes will have an edge angle of from 5° to 7° per inch which is caused by the extreme temperatures generated during the cutting process. 20000° to 60000°.

DROSS REMOVAL

Dross is readily removed with the conventional slagging hammer (Figures 4 & 5) and/or hand grinders. Slag generally flies off in long sections when shocked by hammer. The cut edge of material tends to be harder than the base material but tests have proven that the H.A.Z. of mild steel - T-1 and Van-80 materials in sizes ¼" (6.35 mm) through 3/4" (19.05 mm) will not exceed .040" (1.02 mm) (Figure 6).

NOISE FACTORS

The noise factor is one that must be given much thought considering the current OSHA standards and the ever present possibility that they may go even lower.

Decibel levels will vary with torch ratings and tip size but it is almost a certainty that the running of several plasma torches, combined with other sounds related to plate handling will exceed the allowable 90 DBA.

The addition of water mufflers to the torch heads will greatly reduce the noise level but there are drawbacks to this approach that must be noted.

1. Water mufflers require a water supply of 18 gallons per minute per torch.
2. It is virtually impossible to control the tip standoff since the operator cannot see through the water muffler.

NOTE: It is the writers opinion that to date, no manufacturer has made an acceptable height controller that will work with the plasma cutting speed. Hence, tip height must be controlled manually.

AIR POLLUTION

Both gas and plasma cutting produce smoke that must be exhausted or controlled in some manner. We decided on the water table approach. When water levels are maintained to within ¼" (6.35 mm) to ½" (12.7 mm) to the bottom of plate surface, the scrubbing effect is such that steam is generated during the cutting cycle and very little smoke escapes. This has proven so successful that all cutting within our facilities is now done over water. We maintain two separate and distinct water table systems. One over a self-flushing water bed and one consisting of individual moving water tables. (Figure 7 depicts the latter)

MACHINE UTILIZATION

This is determined by many factors ranging from the skill of the operator through the adequacy and reliability of the material handling system. Torch monitors were installed on each cutting head and machine and torch times (actual firing time per torch) were monitored for a period of four months. The results were that machine run time averaged 72% while average torch time was 58%. While this is considered an acceptable level of utilization, we are constantly searching for ways to improve the system.

CONCLUSION

The plasma method of cutting production parts will result in significant savings in labor and related dollar cost per piece part as compared to conventional machining and Oxy-Acetylene cutting.

Care must be taken when committing parts for this process and only parts that have subsequent weld operations or those that would be acceptable from a quality standpoint (considering the edge taper) should be selected.

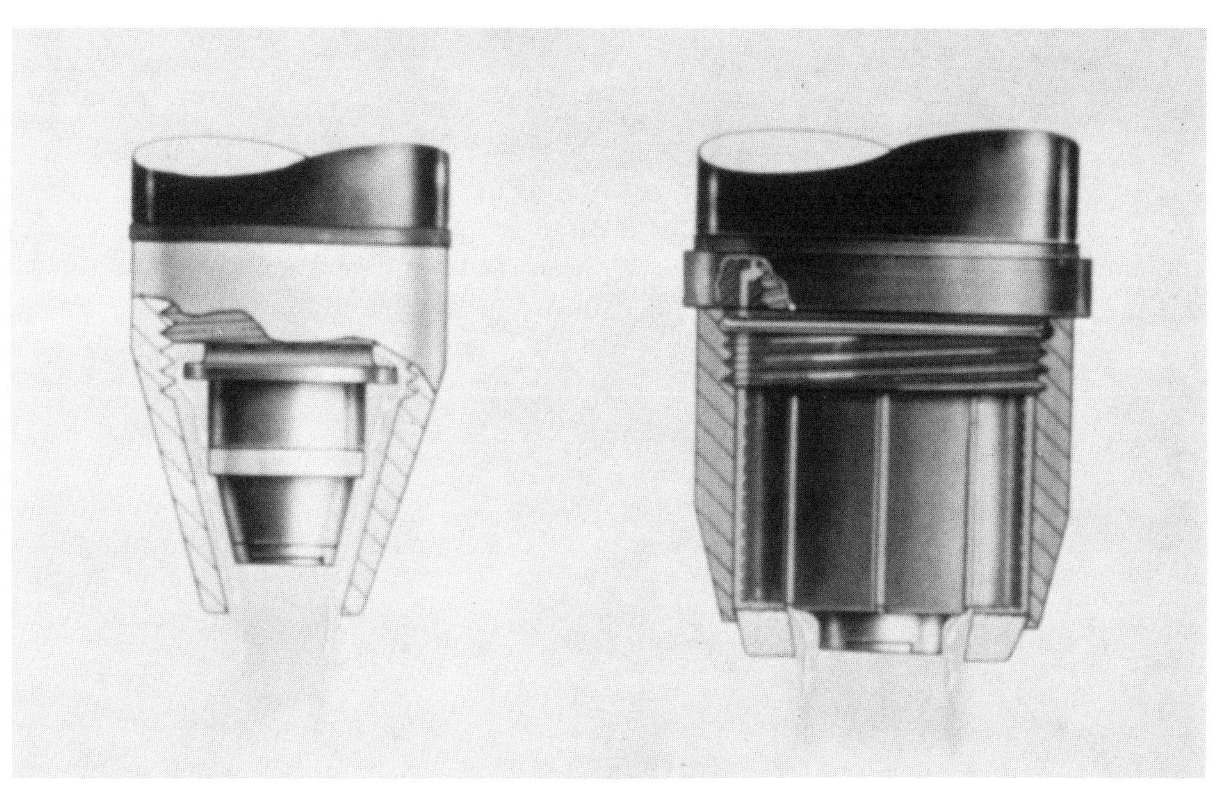

FIGURE 1

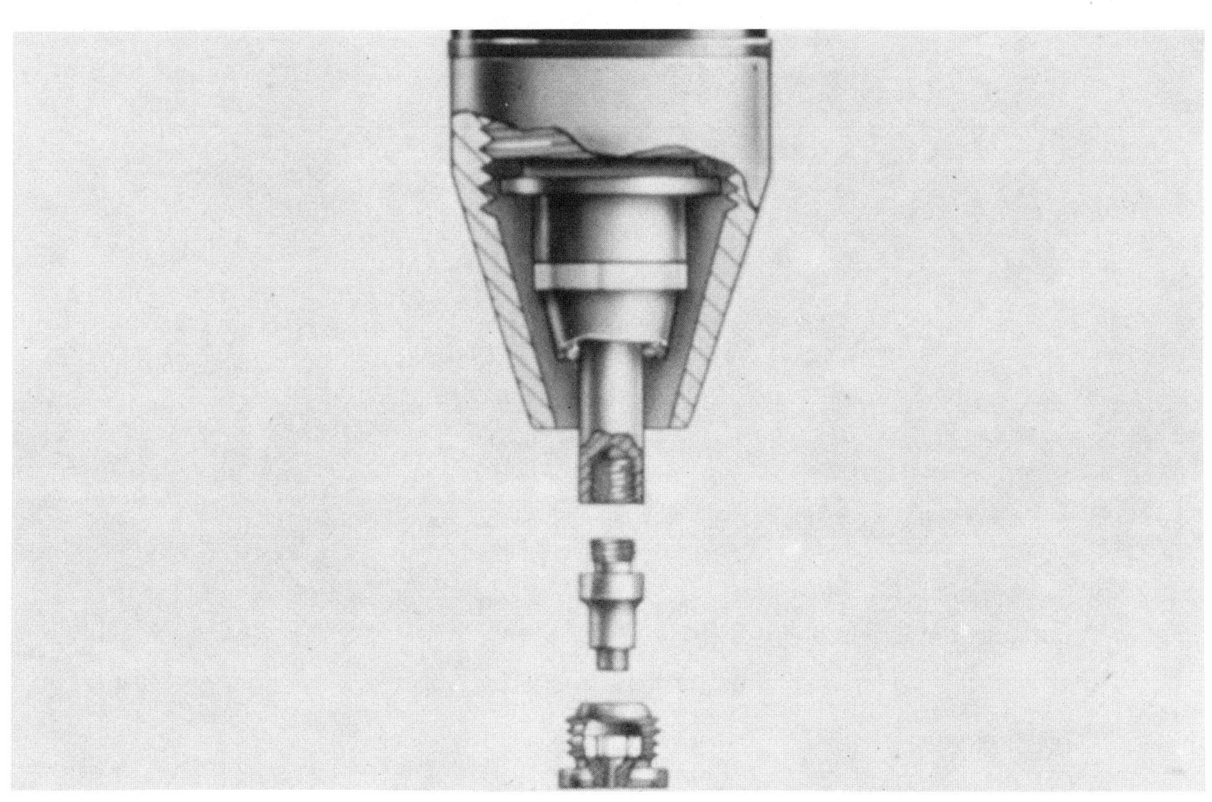

FIGURE 2

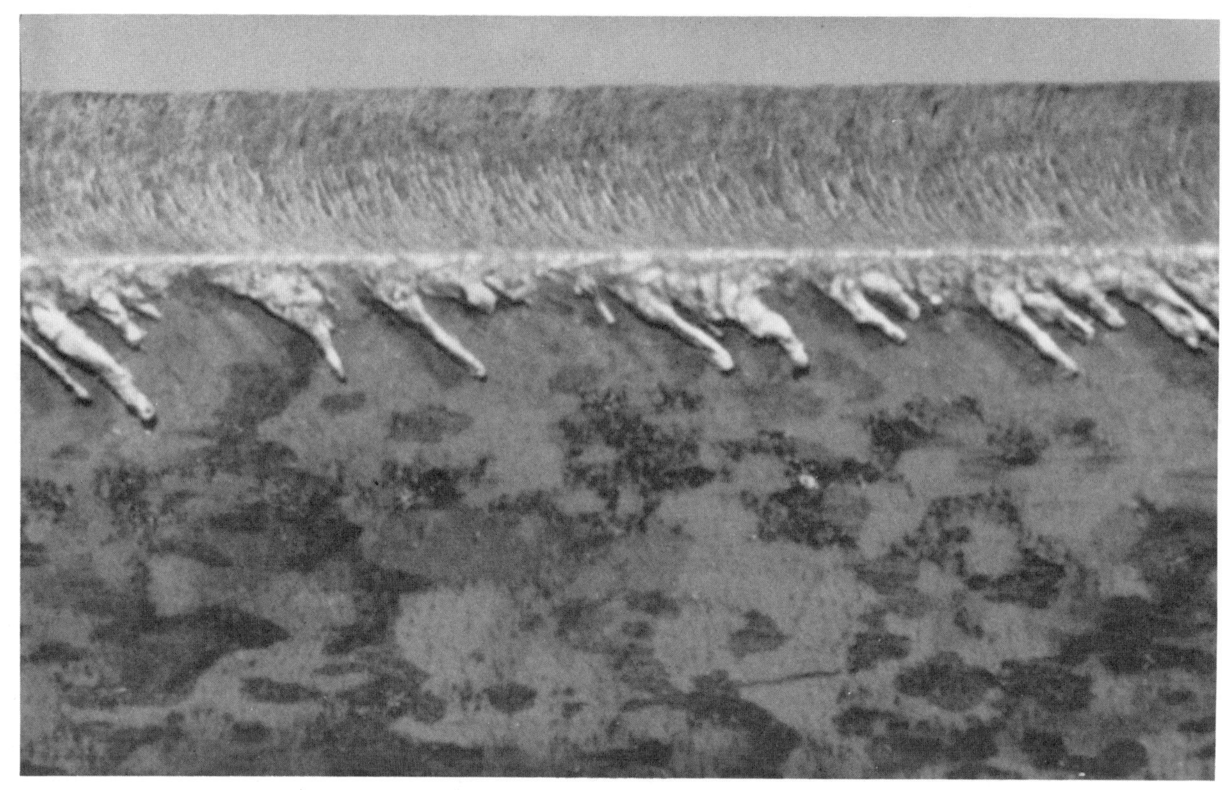

FIGURE 3

FIGURE 4

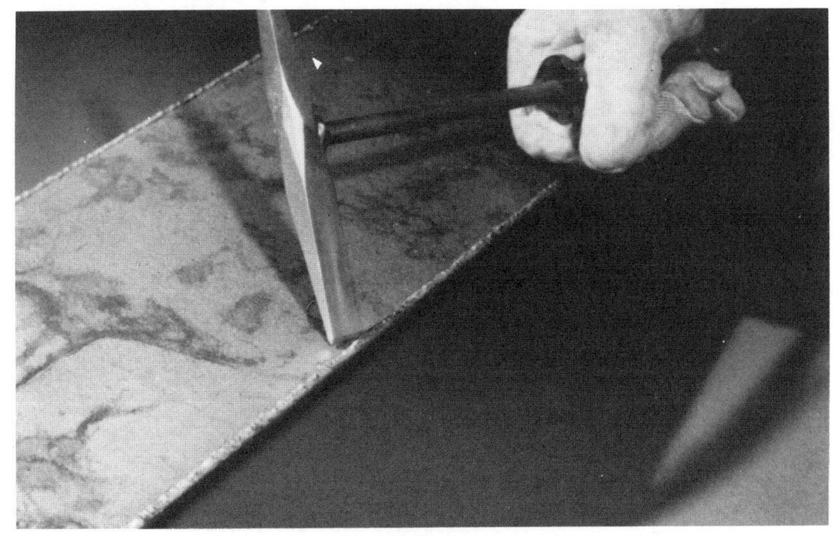

Figure 5

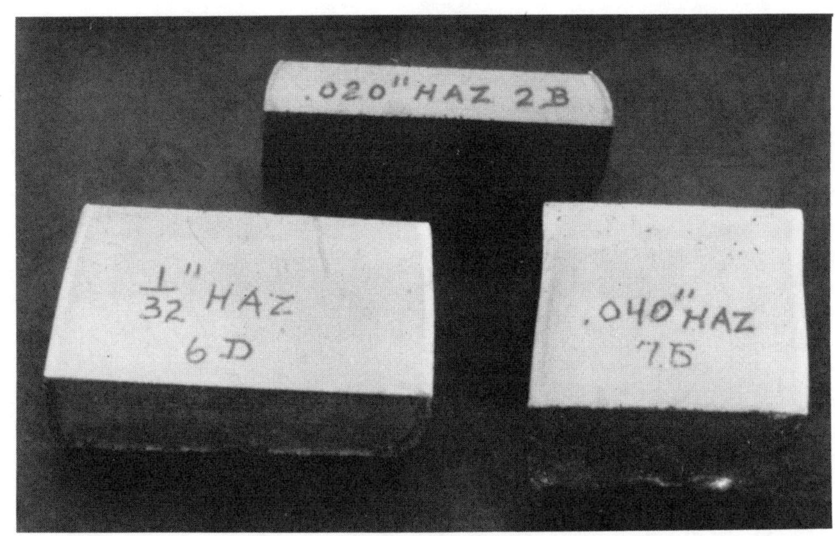

Figure 6

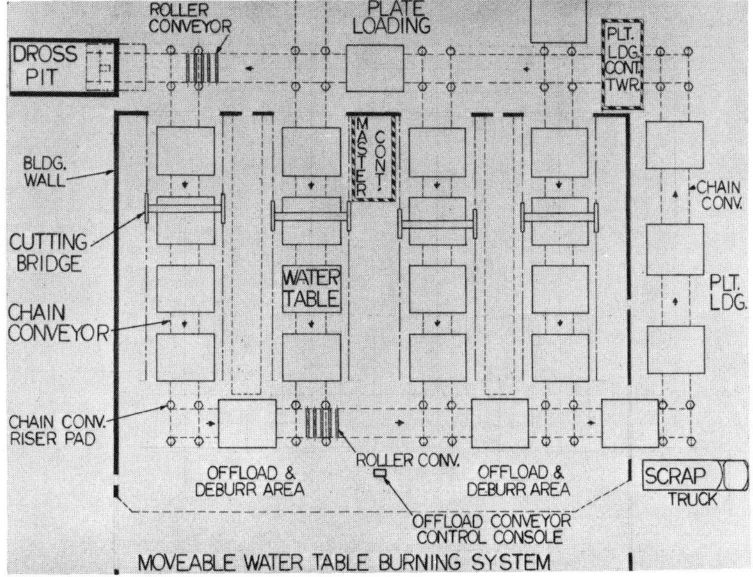

Figure 7 Split axis laser cutting system operated by numerical control.

CHAPTER 4

EDM

Reprinted from: Tooling and Production, March 1977

EDM: how it works

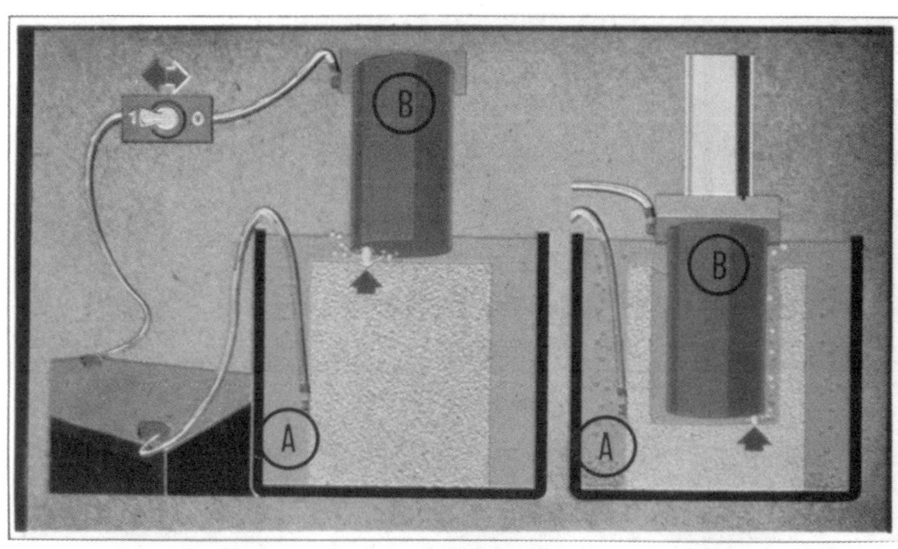

1. The dielectric fluid (A) acts as an insulator between the positively charged electrode (B) and the workpiece. When the two are brought close enough together a spark jumps and bit of metal is removed from the workpiece.

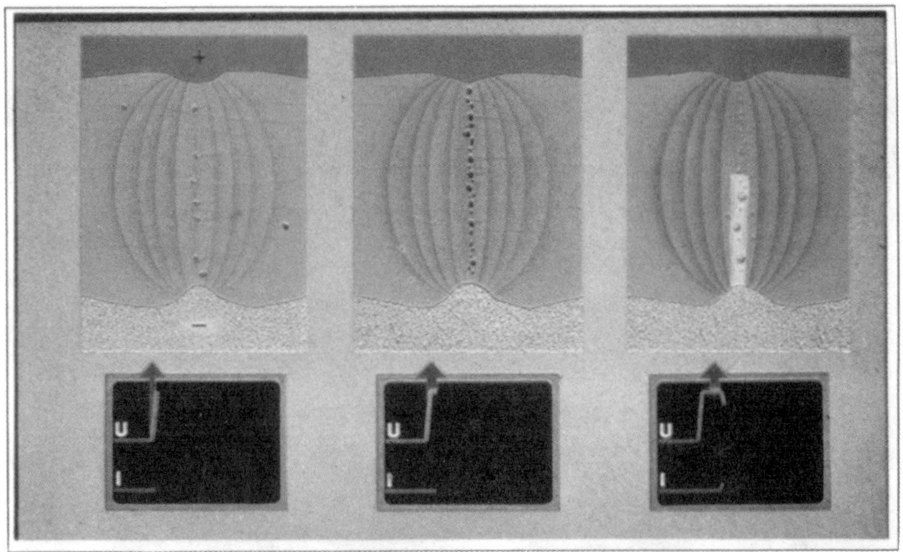

***2.** Though the spark jumps rapidly, we can show by a series of drawings a sequence of things which happen in that short time period. Voltage builds up an electric field and conductive particles in the dielectric concentrate in the field. Collision between negatively charged and neutral particles creates a bridge of ionized positively and negatively charged particles between the workpiece and the electrode tool.*

The principle of electrical discharge machining is simple. The workpiece and tool are placed in the working position in such a way that they do not touch each other. They are separated by a gap which is filled with an insulating fluid. The cutting process takes place in a tank, and the workpiece and tool are connected to a DC source via a cable. There is a switch in one lead. When this is closed, an electrical potential is applied between the workpiece and tool. At first no current flows because the dielectric between the workpiece and tool is an insulator. However, when the gap is reduced a spark jumps across it, **Figure 1**.

In this process current is converted into heat. The surface of the material is very strongly heated in the area of the discharge channel. If the flow of current is interrupted the discharge channel collapses very quickly. Consequently, the molten metal on the surface of the material evaporates explosively and takes liquid material with it down to a certain depth. A small crater is formed. If one discharge is followed by another, new craters are formed next to the previous ones and the workpiece surface is constantly eroded.

The voltage applied between the electrode and workpiece and the discharge current have a time sequence which is shown under the illustrations of the individual phases, **Figure 2**. Starting from the left, the voltage builds up an electric field throughout the space between the electrodes. As a result of the power of the field and the geometrical characteristics of the surfaces, conductive particles suspended in the fluid concentrate at the point where the field is strongest. This results in a bridge being formed, as can be seen in the center of the picture.

At the same time, negatively charged particles are emitted from the negatively charged electrode. They collide with neutral particles in the space between the electrodes and are split. Thus, positively and negatively charged particles are formed. The process spreads at an explosive rate and is known as impact ionization. This is encouraged by bridges of conductive particles.

Here again we see illustrated what in fact is invisible. The positively charged particles migrate to the negative electrode, and the negative particles go to positive. An electric current flows. This current increases to a maximum, and the temperature and pressure increase further. The bubble of vapor expands, as can be seen in the far right of **Figure 3**.

The model shows how the supply of heat is reduced by a drop in the current, **Figure 4**. The number of electrically charged particles declines rapidly, and the pressure collapses together with the discharge channel. The overheated molten metal evaporates explosively, taking molten material with it. The vapor bubble then also collapses, and metal particles and breakdown products from the working fluid remain as residue. These are mainly graphite and gas.

By means of the model we will now try to demonstrate the relationship between the flow of current and heat. In a detailed enlargement we see the negative workpiece surface, and above it a part of the discharge channel, **Figure 5**. Positively charged particles stroke the surface of the metal. These are shown in red. They impart strong vibrations to particles of metal, which correspond to a rise in temperature. When a sufficient velocity is reached, particles of metal, which are shown in gray and yellow here, can be torn out. A combination of positively charged particles, which are shown in red, and negatively charged particles, which are shown in blue, augments the vibration and thus raises the temperature of the particles, which are now uncharged.

We know that electrical energy is converted into heat when the discharge takes place. This maintains the discharge channel, leads to the formation of discharge craters on the electrodes and raises the temperature of the dielectric.

Now let us examine the question of polarity. The exchange of negatively and positively charged particles, which are respectively shown in blue or red, results in a flow of current in the discharge channel. The particles thus generate heat which causes the metal to melt. With a very short pulse duration, more negative than positive particles are in motion. The more particles of one kind move toward the workpiece, the more heat is generated on it. It is also important that, as a result of their greater size, the positively charged particles generate more heat with the same impact velocity.

In order to minimize the material removal or wear on the tool electrode,

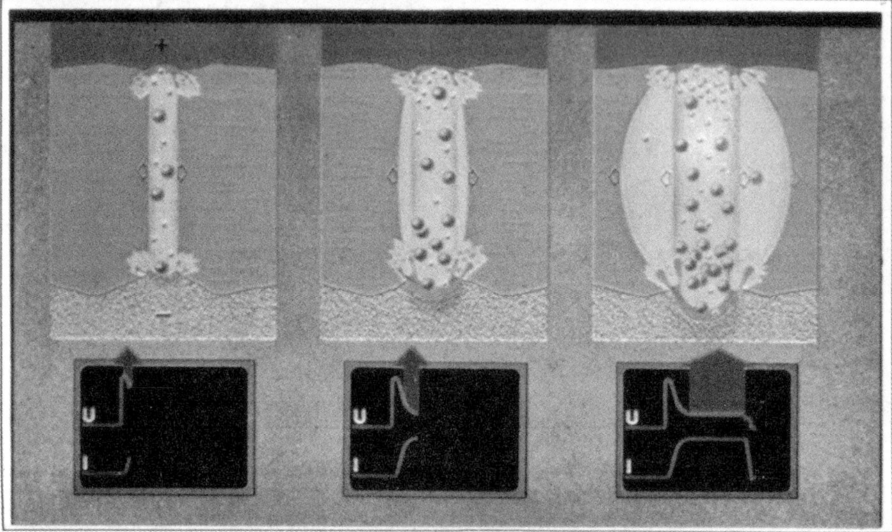

3. Since unlike charges attract one another, the negative ions go toward the electrode and the positive to the workpiece. As current increases, heat rises and these particles form an expanding bubble of vapor.

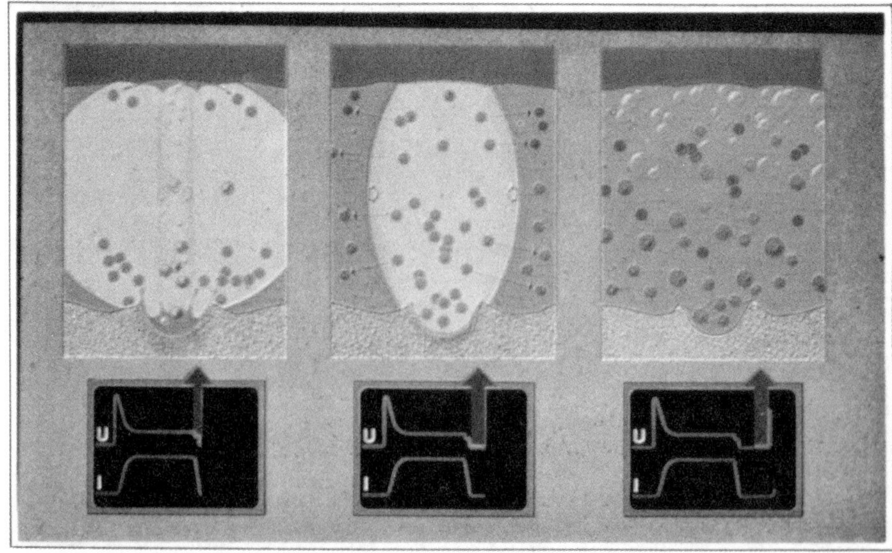

4. The heat in the bridge area causes a small amount of the workpiece to become molten. The vapor bubble then collapses and the metal particles are trapped in the fluid.

the polarity is selected so that as much heat as possible is liberated on the workpiece by the time the discharge comes to an end. With short pulses the tool electrode is therefore connected to the negative pole. Its polarity is thus negative. With long pulses, however, it is connected to the positive pole so that its polarity is positive. The pulse duration at which the polarity is changed depends upon a number of factors which are mainly connected with physical characteristics of the tool and electrode materials. When steel is cut with copper, the marginal pulse duration is about 5 microseconds, **Figure 6**.

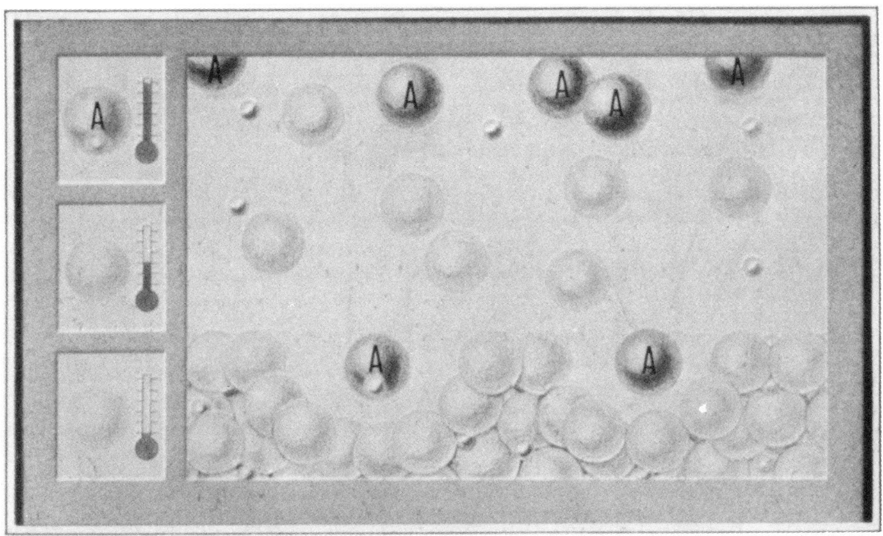

5. Positive ions (A) bombard the surface of the workpiece, which causes the metal to vibrate and rise in temperature. Metal particles break free and those which are negatively charged combine with the positive ions to produce neutral particles. Read the pictures on the left from top to bottom.

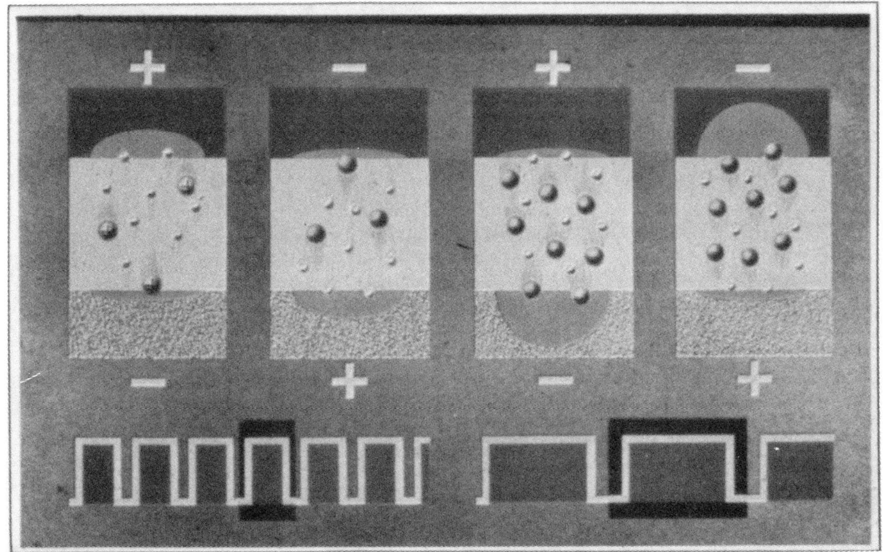

***6.** If the duration of the pulse is short, more negative than positive ions are in motion. Since we want maximum wear on the workpiece rather than the electrode, this is a good time for the red electrode tool to be negative. This situation is shown by the three drawings on the left half of this figure. If the duration of the pulse is long, more positive than negative ions are in motion and it is advantageous for the red electrode to be negative.*

The power-supply unit is an important part of any spark-erosion system. It transforms the AC supply from the mains and provides rectangular voltage pulses. This can be visualized by plotting a graph of voltage against time. By a number of switching devices, the size of the rectangles and the distance between them can be adapted to any operational requirements.

Erosion with a light current gives a low rate of removal, while conversely, a heavy current gives a high rate of removal. But the wear on the electrode expressed as a percentage of the volume also increases if steel workpieces are eroded with copper electrodes. Graphite electrodes behave differently. The wear declines up to a certain current level and then remains more or less constant.

Eroding with short pulses means increasing electrode wear. Conversely, the wear is smaller when the pulses are long. In practice, when roughing with copper and graphite electrodes into steel, a pulse length lying between maximum removal and minimum wear is selected.

The interval between two discharges is a factor of considerable importance. In general, we can say that rapid removal with little wear can be achieved with small intervals or, in other words, a high duty factor. The limit must not be exceeded because a point is then reached beyond which the process is impaired, resulting in reduced erosion and greater wear. This critical value is also known as the marginal duty factor.

A rougher surface is machined to a finer one by eroding with reduced discharge energy. The roughness is reduced, while the electrode wear is somewhat increased.

In workshop practice, in roughing or premachining a degree of roughness should be attained which needs only to be evened out in the next machining stage. Experience has shown that the roughness of the subsequent stage is about a third to a fifth of the initial roughness. This procedure gives a very economical overall eroding time in relation to the degree of accuracy attained.

EDM And The Die Designer

By Richard T. Benger
President, Prospect Tool and Die Company

1. WHAT EDM MEANS TO THE DIE DESIGNER

   An approach to the questions most frequently faced by the die designer.

2. CHOICE OF ELECTRODE MATERIALS

   How to select the proper material for electrode fabrication for different stamping die applications.

3. BLANKING DIES

   Various methods are discussed for producing a variety of blanking dies including techniques for controlling punch to die clearance.

4. PROGRESSIVE DIES

   This section features the manufacture of a progressive die with many narrow slots, and describes the methods of set up or location of punches to pilots.

5. CARBIDE DIES

   Rigidity is the keynote to high production runs from carbide dies. The methods described have been successful when using EDM for carbide die construction.

6. CONCLUSION

WHAT EDM MEANS TO THE DIE DESIGNER

The use of electrical discharge machining in the manufacture of stamping dies begins with the die designer. Complex shapes can be designed from solid die blocks as opposed to the previous method of "splitting" a die block, which will minimize bench time at assembly. The fit of punch to die or "die clearance" is uniform around the entire contour of the punch as the overcut of the electrode is controllable by the amount of power (amps) from the power supply. The amount of the die clearance is controlled by the length of the spark discharge. Clearance from .0001 up can be specified by the designer to meet stock thickness requirements. Different die designs will require different approaches for the EDM process and the designer can specify whether a die will be cut from the die face or from the back of the die block. Compound dies may be designed to be cut from the die face as the blank does not have to pass through the die whereas in a blanking die, blank clearance or taper must be provided for ease of blank discharge from the die.

No longer is the designer tied to sectional dies for weak or troublesome areas. If a solid die block does fail, the immediate area may be trepanned out with a round, square or other convenient shaped electrode and a section may be fitted into the cavity. Electrodes for this purpose may be stock sizes of brass or copper tubing. Using the same techniques, engineering changes may be accomplished in one piece dies.

Recognizing the possibilities of such failures, location of screws and dowels must be kept clear of the area. This is the responsibility of the Die Designer!

When long runs are specified or when the material to be punched is abrasive or hard, then the designer must consider tool cost and tool maintenance versus the number of parts produced per grind of the die. Generally speaking,production will be 100% to 1,000% greater with carbides.

CHOICE OF ELECTRODE MATERIALS

The first choice for electrodes for a general class of die work is graphite, when using a tube power supply. This material is economical to purchase, easy to machine and has a relatively good wear ratio. Wear ratio is expressed as units of electrode wear to units of work piece cut. Example: a good grade of graphite will have a wear ratio of 2½ to 1. A 1.00 die block will consume approximately .400" of graphite to EDM a contour through it. Because grades of graphite will vary it is advisable to allow a safety factor of 50%. Copper-brass, copper-graphite and copper-tungsten are also used for electrodes for stamping dies. Very thin sections will require the rigidity of copper tungsten to withstand the pressure of machining the electrode. Brass or copper may be used when acid etching is required to control punch to die clearance. Copper tungsten or the best grades of copper graphite must be used when cutting carbide die blocks or punches.

When a transistor power supply is used, copper will give excellent wear, however the EDMing time is increased by 100%. It is sometimes advisable to produce a 1/8" thick copper electrode with a 2D pantograph which will, with a shank attached, be enough to EDM the die block, and female electrode.

When the die block is EDM'ed from the back, a shank or holder must be attached to the electrode so that it is long enough to reach through the die block.

Small electrodes may be mounted to a shank with fast drying cement, however, care must be taken to insure using only a very thin coating, as the cement is non-conductive. When the coating is heavy it will form an insulation which will not allow the current to pass, which in turn will result in a damaged electrode.

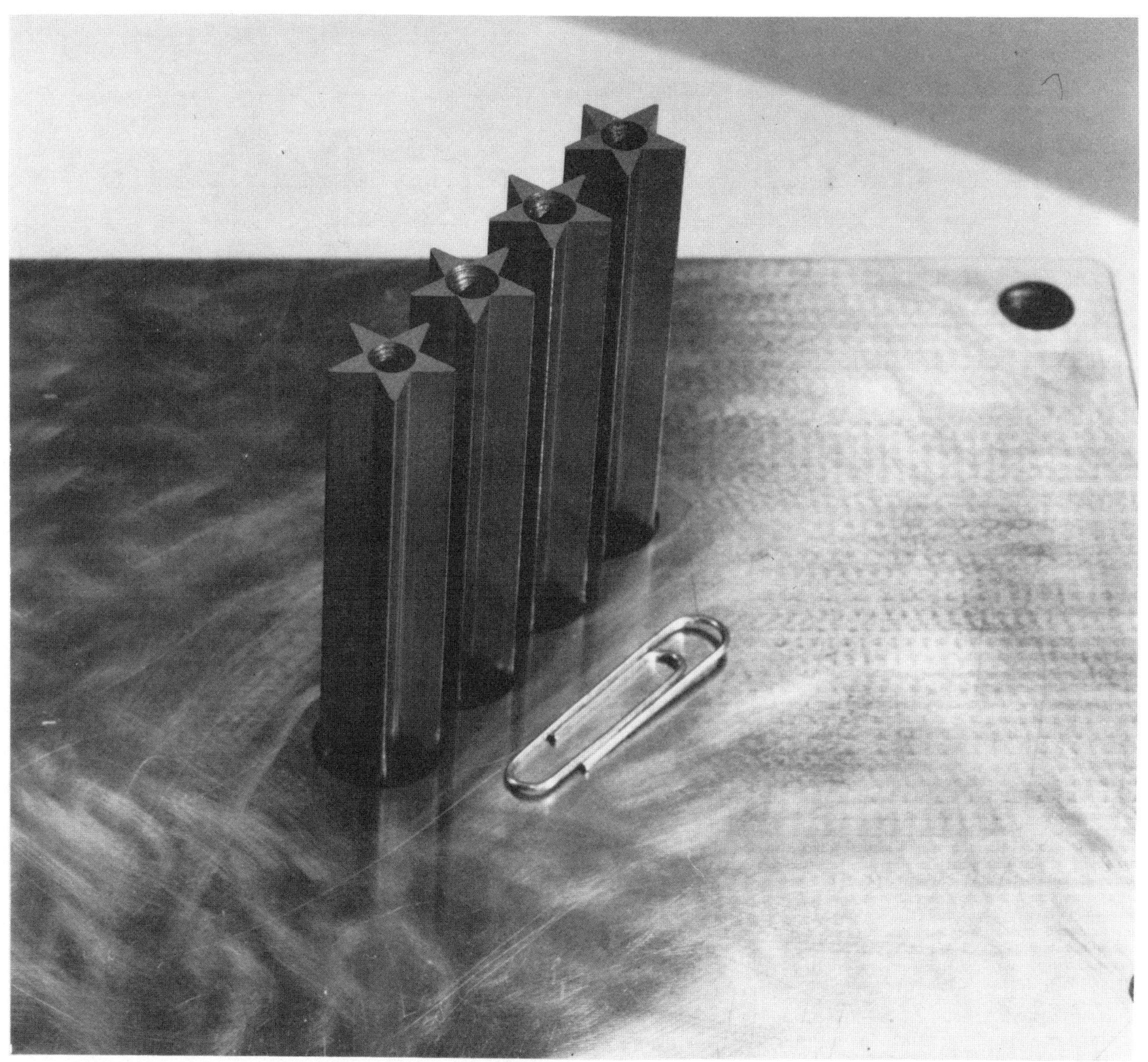

Fig. #1
Four graphite electrodes ground and mounted in the permanent punch holder. Note the length of the electrodes is long enough to cut the stripper die plates and female electrode for this high production blanking die for gummed stars. When the die sections are weak or thin more than one die plate may be EDM'ed and held for future use. If the cross section is fairly rigid, the electrodes may be made long enough to produce a second die plate at a future date.

BLANKING DIES

One piece blanking dies offer many advantages for high production stamping, among which are ease of fabrication, uniform punch to die clearance, low maintenance and higher production runs.

If the blank to be stamped is large enough, the electrode blank may be mounted on the face of the punch with screws and the grinding of the punch (after hardening) also produces the electrode. The die block may be EDM'ed from the top with the punch and die block mounted in an insulated bushing die set. Care must be exercised when EDM'ing from the top as the EDM process will generate some front taper in the die block.

When front cut dies are run in a punch press the punch must pass completely through the die block so that no jamming or wedging of the blank in the die will occur. The taper generated (app. .003/inch) will not affect the die or blank if the blanks are not allowed to accumulate in the die.

The stripper and the cavity in the die set may also be EDM'ed with the same electrode, providing it is long enough. The average die may be EDM'ed with a 10 ampere cut at 125M cycles and the stripper and die set can be cut with a 30 to 40 ampere cut at a lower frequency which will produce a larger cavity in the die shoe for blank drop out. The physical size of the electrode will determine the amount of amperes an electrode can draw, for instance, an .060 diameter electrode may be able to draw only 3 amperes whereas a 1.000 diameter electrode may be able to draw 50 amperes, depending upon the particular power supply being used.

If greater clearance is desired in the die shoe the cut can be first made directly under the die cavity, then the electrode may be moved off center and a second or third cut may be made to enlarge the die shoe cavity. In most instances the heavier EDM cut is adequate because we can easily produce .007/side overcut if finish is not critical.

The inside contour of the die block should be rough machined to within 1/32" of the finished contour before heat treating. An excellent method for EDM'ing the contour is to EDM a .003 deep cut in the soft die block from the finished electrode as a guide for band sawing or rough machining the cavity.

If the punch cannot be ground through, a different approach is used to produce the punch and die cavity. First a male electrode is machined to the dimensions of the die cavity less the determined overcut, then the prehardened die block is EDM'ed either in or out of the die shoe. A female electrode is also prepared similar in contour to the die block, which is also EDM'ed with the original male electrode. This will then EDM the punch contour by reversing the polarity of the power supply.

After the finish EDM cut the die block and punch must be retempered to remove stress in the EDM finish. Insofar as the process is a burning or melting process the heat generated is extremely high and the surface is automatically quenched at the end of the EDM cycle in the dielectric bath. The tempering should be done at the same temperature as the original heat treat temper thereby assuring that the die block will not move or change shape. This is most critical when punch to die clearance is .0005 or less.

A steel or brass blank .030 thick will require .0015/.002 clearance between punch and die and this clearance will be obtained with the setting of the power supply. If the electrode is ground with the punch, the "overcut" will generate the correct clearance with no further work on the electrode. If the application requires a clearance of .0005 and the electrode is ground with the punch, then the electrode should be brass or copper and it may be reduced evenly with acid etching techniques.

The times specified are based on a test sample of brass rod 1" diameter 2½" long. All tests were made at room temperature. Caution must be exercised when etching large electrodes as a certain amount of heat is generated and the etching speed will increase. For most uniform results acid solution can be preheated to 120° F. A small workpiece will generate very little heat and times established with a test sample may be used to determine the length of time required to reduce the electrode an exact amount. The amount of reduction is usually the amount of overcut that must be allowed to obtain an exact contour of the original electrode in the finished die plate or workpiece.

The electrode must be free of all oil, grease, and dirt and must be kept in motion during the etching process to prevent the formation of gas pockets which will cause an uneven etch. A successful method of preventing gas bubbles is to continually scrub the electrode with a toothbrush during the etching process. As the solution becomes saturated through use, the time required for a given amount of reduction will increase. When the time doubles, replenish the solution. A 30 second wash in running water is adequate to check the electrode for dimension. A base solution of baking soda and water may be used to completely neutralize the acid action after a water rinse.

The following examples will give excellent control of electrode size using acid mixtures.

CAUTION: ALWAYS ADD ACID TO WATER

NITRIC ACID 90°

Mixture No. 1

2 parts water and 1 part acid

.0009 was removed from surface during first minute

.0053 was removed in five minutes

Mixture No. 2

3 parts water and 1 part acid

.0003 was removed during first minute

.0018 was removed in five minutes

Mixture No. 3

4 parts water and 1 part acid

.0001 was removed during first minute

.0005 was removed in five minutes

If the contour of the blank is uniform about a center line such as a clock hand, the die block may be split and the two die sections with two female electrode sections may be ground as a sandwich thereby producing, with one grinding setup, the finished die block and the female electrode which will be used to EDM the punch. A male electrode may then be made from the female that will be used to EDM the stripper plate.

When heavier gauges of stock are to be punched the die block will be thicker and must be cut from the back. If the electrode is large enough, the power supply is set to draw 30 to 40 amps. This will produce approximately .005 overcut/side. When the electrode is within ¼" of the die face the power is reduced to .0015 overcut for the balance of the cut. The blank or slug will now drop from the die freely after it passes the ½" level.

Fig. #2

This two cavity pierce and blanking die was cut using the male to female technique. Male electrodes were machined to product print dimensions minus predetermined overcut. Die cutouts and female electrodes were cut from the males and punches were cut from the females.

PROGRESSIVE DIES

Most progressive dies are cut from the die face with electrodes mounted to the face of the punches. All electrodes must be made from the same material to obtain a uniform cutting rate and punch to die clearance. When punching narrow slots the electrodes may be tellerium copper and attached to the punch blank by electron beam welding.

Grinding of the punch and electrode is then accomplished as one operation. This process will anneal the punch for only .020 which must be ground off when the electrode is removed.

All profile punches may be set up on the punch holder, after grinding, and located to the pilot piercing punches and pilots.

If the clearance is to be .0001 to .001, the electrodes must be reduced by etching (see etching under blanking dies). An impression is then EDM'ed .003 deep on the face of the soft die blocks which are located to the pilot punches and pilots by conventional jig boring or master dowels and jig grinding. Narrow slots may be rough machined by drilling a series of round holes. Larger areas by milling or band sawing to within 1/32" off the EDM impression. All electrodes must be the same height in order to create a seal for oil pressure which is pumped through a manifold under 2 to 15 PSI.

Holes in the die block that are not being cut by EDM, such as round pierced holes or pilot holes must be plugged during the EDM process to prevent loss of oil pressure. This may be accomplished by wood dowels driven into the holes or magnets placed over the holes.

The manifold under the die must be large enough to encompass the area of all the cavities to be cut by EDM to insure oil pressure to all cavities.

After the hardened die is EDM'ed the stripper is positioned and EDM'ed with the same electrodes. The die shoe is also marked with the electrode (.003 deep) and the cavities are machined by conventional methods. After the electrodes are removed every profile punch will have the identical clearance around the entire contour! Forming stations are ground and mounted separately on the die shoe.

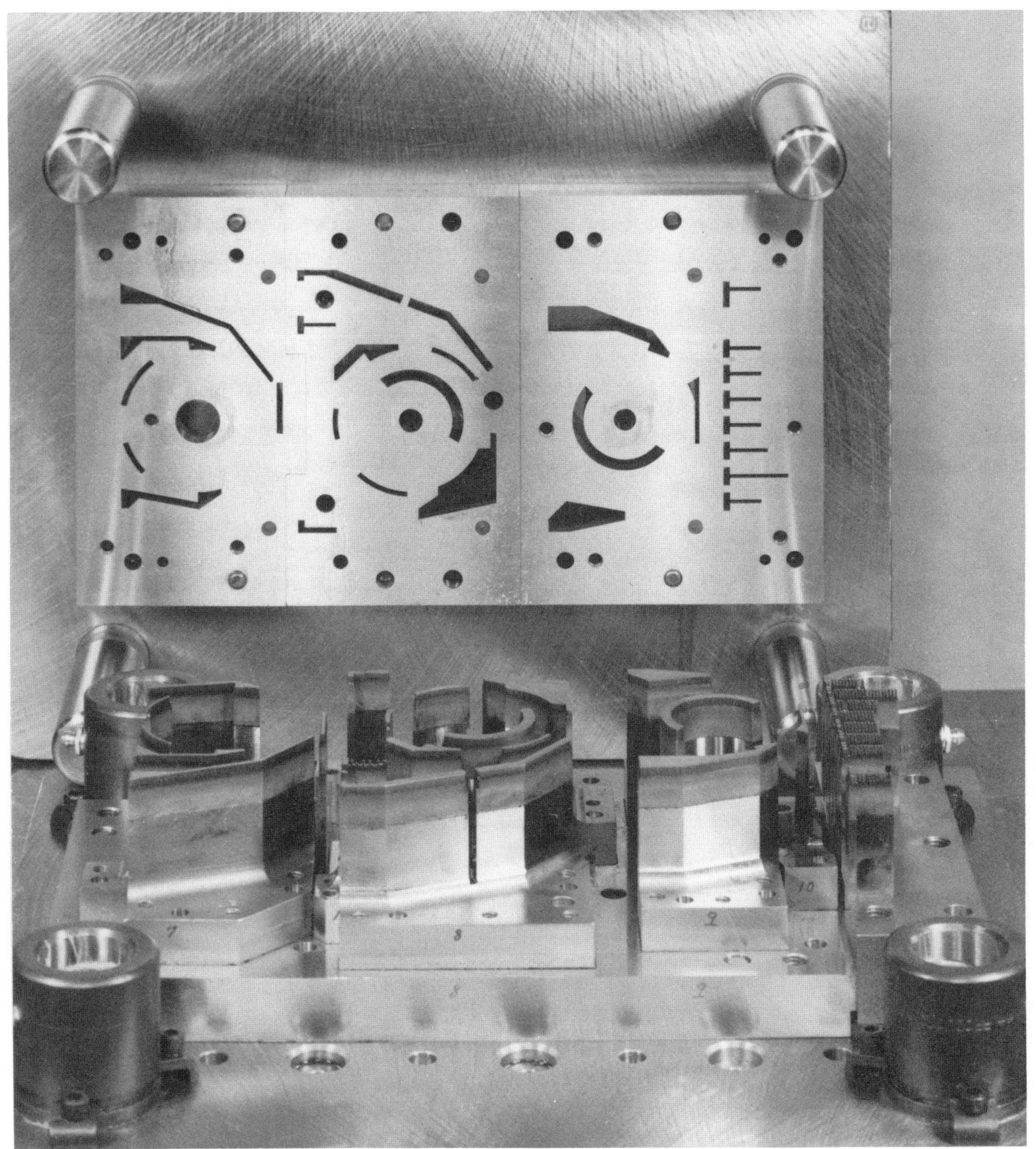

Fig. #3
A progressive die to punch printed circuit copper overlay .012 thick features punches tipped with tellerium copper and electron beam welded to HCHC prehardened punch blanks. The die plate is in 3 sections to facilitate manufacture and die cutouts were premachined from an electrode impression .003 deep. As described, same die cavities were premachined by drilling a series of holes. EDM'ing was done in an insulated die set thereby assuring uniform clearance.

CARBIDE DIES

The die designer must be aware of the virtues of carbide for the selection of the proper grade, estimated production runs and maintenance procedures.

Experience has shown that four post die sets are a must for high production runs from carbide dies. Carbide die sections should be supported in a hardened case or support base to prevent transverse rupture or fracture of the carbide. All steel die sets are good insurance against deflection of support members which could cause failure.

When possible, punches should be "heeled" with a lead guide on the heel portion to prevent possible failure if half blanks are punched. It is not necessary to retemper carbide die sections as no stresses are created by the EDM process.

The accepted grade of carbide for blanking dies is C-13 however, grades C-10 or C-12 should be used when punching heavier gauges of stock or when the stock is tough or hard.

The surface of carbide would be lapped after grinding to obtain maximum production.

Example: Material 1/16 Mica

HCHC Die 150,000 pcs per grind

Carbide die ground not lapped 400,000 pcs per grind

Carbide die ground and lapped 750,000 pcs per grind

Difficult situations with delicate sections may require the use of carbide for die sections and either HCHC or high speed steel for the punches.

CONCLUSION

During the last fifteen years the EDM process has revolutionized stamping die design when using both HCHC and carbide. Fewer skilled man hours are required to "bench fit" a die after being cut by EDM. The typical die shop can produce longer running dies due to uniform punch to die clearance and the use of carbides.

The need to "shave" a punch or die section is eliminated when both male and female electrodes are used to control size and the fit of punch to die. This process is most useful for compound dies when the shedder must fit the die cavity with a slide fit. A typical example is a compound die to stamp a brass gear .010 thick with a splined center hole.

The transistor power supply allows the die maker to work with minimum thickness electrodes from telerium copper. Complex electrodes should be considered for this approach. Copper electrodes may be electrochemical milled and stacked for cutting the die plate and female electrode and the die designer must be aware of all the approaches of EDM for todays stamping requirements.

Reprinted from: Manufacturing Engineering, July 1976

Refeed: Key to Production EDM

Development of the refeed unit enables an EDM system to put an electrode at precisely the correct position at the beginning of a machining cycle – and therein lies the secret of production EDM

DANIEL B. DALLAS
Editorial Director

1. TRANSFER LINE FOR EDM is seen in this photo. Refeed principle is key to this machine's design.

EDM AS A TOOLROOM TECHNIQUE is sufficiently well established to be accepted as a traditional rather than nontraditional methodology. Equally well established is the transfer line. What is new, and far from traditional, is the concept of a transfer machine that utilizes EDM at all machining stations. Such a line has been developed and is now in the final tryout stage at Raycon Corp., an Ann Arbor pioneer in the use of EDM for small hole drilling. As in most other Raycon applications, the transfer line — shown in *Figure* 1 — is for hole machining.

Transfer Line Design. The line, a nonsynchronous unit, comprises an automatic feed system, a rough drilling station that drills eight holes half way through four parts; a second rough drilling station that completes the hole drilling operation; a third station that EDM reams the holes to tolerances under 0.0001 inch (0.003 mm); and a fourth station that EDM machines a metering edge on the parts. The "metering" edge is comparable to a shallow counterbore. Distance between the metering edges is extremely critical. Accordingly, the repeatability of the EDM process comes into its own in this application.

The parts machined in this line are components of a diesel fuel metering system. They're hardened steel, approximately the size and shape of a filter cigaret. Four units are automatically loaded into a pallet at the beginning of the machining sequence. Pallet then advances through the various machining stations. Upon arrival at each station the pallet is hydraulically retracted into position for the related machining operation. Control of all transfer line functions is exercised by a Modicon programmable controller. Exact production volume — which is privileged information — is in excess of two hundred parts per hour.

The Refeed Operation. Key to the success of this line — and of all the production EDM units built by Raycon — is a patented electrode feed unit called a "refeeder." The problem solved by the refeed unit is that of precise positioning of the electrode prior to the beginning of a machining operation. As in any machining operation, correct positioning of the tooling — in this case the electrode — is mandatory. This problem is compounded in EDM by the relatively fast electrode erosion that occurs during machining. This factor, more than any other, has inhibited the use of EDM as an effective production tool.

The system designed by Raycon to solve this problem is shown in *Figure* 2. Three movements are involved, i.e., movement of an antishort slide, movement of an EDM carriage, and wire advance through drive rollers. Again, the objective of these multiple movements is to bring the electrode to a predetermined position prior to the start of the machining operation. How this is done can be seen by examining the operational sequence shown in *Figure* 2.

► POSITION 1. The anitshort slide and EDM carriage are in their fully retracted positions. (The wire can be anywhere as long as it is through the wireguide.)

► POSITION 2. The antishort slide advances by an amount equal to the predetermined "antishort stroke." Since the position of the EDM carriage relative to the antishort slide does not change, all other components — the wire reel, rollers and cartridge — advance with the antishort slide.

► POSITION 3. At this point the feed rolls are activated. Their function is to advance the wire until it is in light contact with the workpiece (or in some cases with a reference block). An electrical signal — actually a low-voltage electrical discharge — completes a circuit, thus stopping the motor that drives the feed rolls.

► POSITION 4. The antishort slide now retracts. The amount of retraction — exaggerated in this drawing — is just enough to permit the machining operation to begin. The most important thing to note is that the refeed unit is back at the point it was at in Position 1, but the wire is now in machining position.

► POSITION 5. Machining begins by servo controlled advance of the EDM carriage. The antishort slide maintains its retracted position. Upon completion of the machining operation, the EDM carriage will have retracted to Position 1. Where the tip of the electrode will be depends on the amount it has eroded during machining — but where it will be doesn't really matter. By again going through Positions 2, 3, 4, and 5, the tip will have been returned to correct start position.

The sequence of operations just described implies that only one electrode

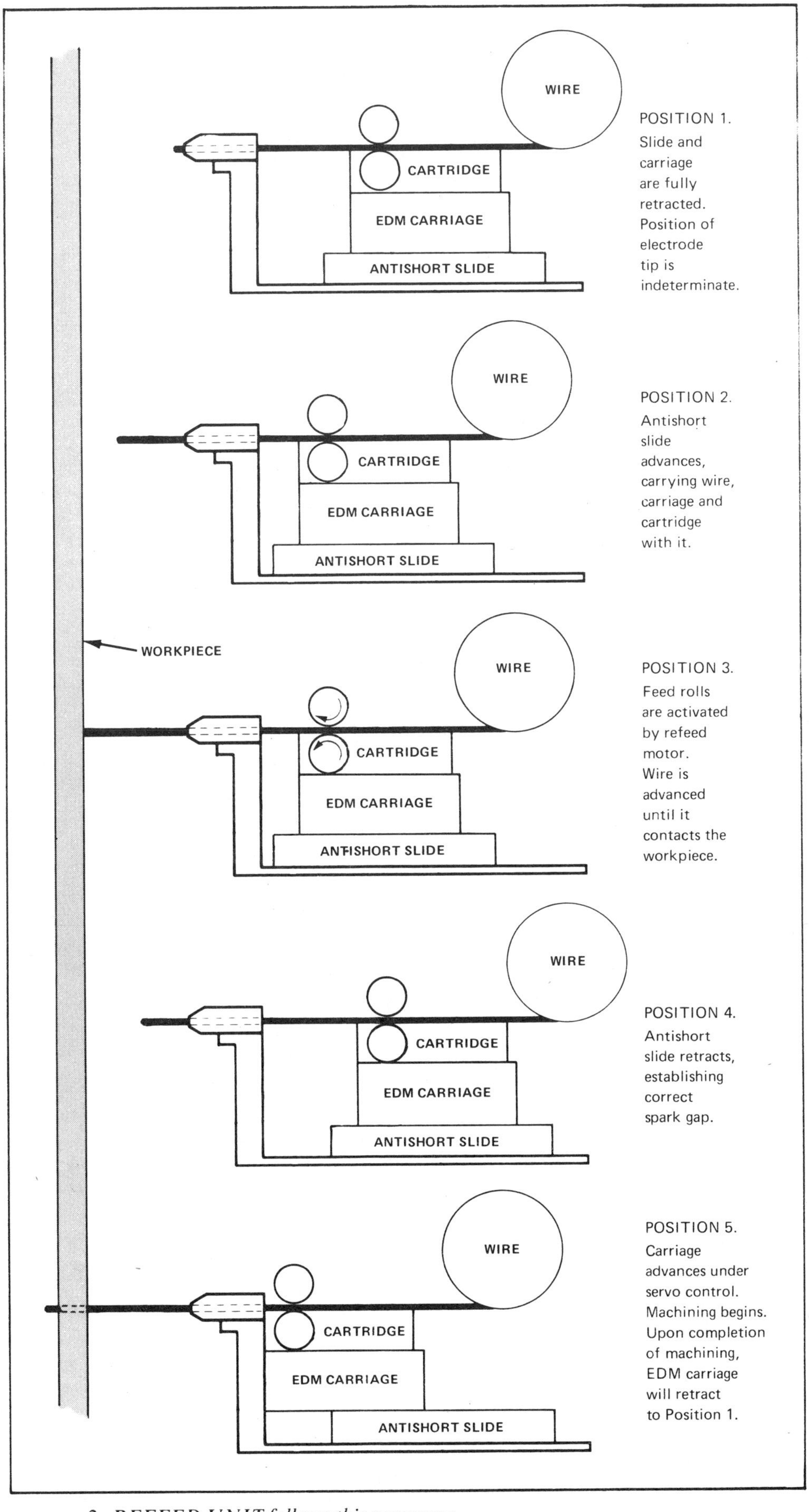
WIRE
CARTRIDGE
EDM CARRIAGE
ANTISHORT SLIDE
POSITION 1.
Slide and carriage are fully retracted. Position of electrode tip is indeterminate.
WIRE
CARTRIDGE
EDM CARRIAGE
ANTISHORT SLIDE
POSITION 2.
Antishort slide advances, carrying wire, carriage and cartridge with it.
WORKPIECE
WIRE
CARTRIDGE
EDM CARRIAGE
ANTISHORT SLIDE
POSITION 3.
Feed rolls are activated by refeed motor. Wire is advanced until it contacts the workpiece.
WIRE
CARTRIDGE
EDM CARRIAGE
ANTISHORT SLIDE
POSITION 4.
Antishort slide retracts, establishing correct spark gap.
WIRE
CARTRIDGE
EDM CARRIAGE
ANTISHORT SLIDE
POSITION 5.
Carriage advances under servo control. Machining begins. Upon completion of machining, EDM carriage will retract to Position 1.

2. REFEED UNIT follows this sequence to establish correct spark gap prior to machining.

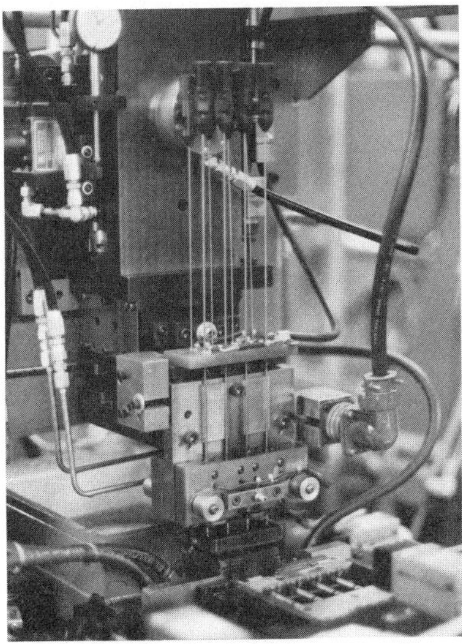

3. INDIVIDUAL EDM STATION in the Raycon transfer line. Drilling of eight holes is performed in this station.

can be preset. In practice, a number of electrodes can be fed to the work through a series of feed rolls, *Figure* 3. A slight amount of slippage in the feed rolls provides for inequalities in starting length when the electrodes contact the work at Position 3.

The sequence also implies that only wire can be used in a refeeding system. Actually, heavier, magazine-fed electrodes can be used. Finally, the electrodes need not be round. The system is adaptable to rectangular, square, and shaped electrodes. However, most of Raycon's work is in the development of machines for hole drilling — applications that almost invariably require wire.

Electrode Dressing. Still another development seen in the Raycon transfer line is a system for dressing the individual electrodes. The refeed is effective in compensating for axial wear, but there's still the problem of radial erosion. To compensate for this, Raycon has developed a dressing system that removes the worn tip prior to refeeding. The dresser is a rotating graphite wheel placed in the horizontal plane. Between machining cycles, the dresser is brought into position, and the electrode is lowered for removal of the eroded tip. It can then be returned to correct start position via the refeed system.

Inclusion of EDM in a transfer line setup does not mean that the technology is now ready to replace drills, reamers, and insert tooling. But it does mean that the process is now well out of its "toolroom-only" category. It is, in every sense of the word, a bona fide production process. ■

Reprinted from: Metal Stamping, June 1977

How to Estimate EDM Time Requirements

By Leif Houman, Eltee Pulsitron, W. Caldwell, N.J.

It is difficult to forecast the exact length of cutting time for EDM work. Most of the time the quotation is made on the basis of an educated guess. This isn't reliable and there is a better way.

Due to the nature of EDM it is difficult to forecast the exact length of cutting time and in turn come up with an accurate quotation for a given job. Most of the time a quotation is made by an educated guess, based on previous similar cuts. This has proven to be a not very reliable method of estimation. In order to help this situation, a more scientific approach has been developed.

1. Choose surface finish and wear
2. Establish metal removal rate
3. Calculate metal removal rate
4. Evaluate flushing mode
5. Ratio of pulsation or interruption of cut to help flushing

One of the first things to establish is the metal removal rate. Some EDM manufacturers do have rather extensive charts showing the metal removal rate. These figures cover a particular electrode material/workpiece material combination cut with one particular dielectric by a specialist under ideal conditions. You might be the one out of a hundred who matches this description. If you are one of the lucky ones, fine, use the charts, but if not then you have to make your own. This sounds right off hand like an enormous task. But looking at it realistically you only have to get the figures for the 3 or 4 workpiece material/electrode combinations at the 3 or 4 roughing and finishing settings you have ended up using. That way you will have these figures cut on your machine with your dielectric and your expertise.

What is the easiest way to obtain this information?

To eliminate the flush problem and other disturbing factors, the electrode for the test cut has to be designed with that in mind. A tubular electrode of the electrode material you want to test with flushing thru the center serves this purpose. An outside diameter of .500 with an ID of .250 will give you a reasonably fast cut for medium to roughing cutting and OD .285 and ID .250 for finishing to medium cutting. (Area of contact of second electrode 1/10 of first electrode). This electrode is cut into the workpiece material you want to test to a depth of .100 inch, at whatever setting you have choosen. The time for the cut should be recorded in minutes. Looking at the left vertical line of the metal removal chart, you find the proper horizontal line representing the number of minutes the cut took. Where that line hits the curve, you go straight down to the bottom line which shows the metal removal rate in cubic inches per hour. If you have used the small electrode the metal removal rate should be divided by 10 as the area of contact is one tenth as small as the big one. This procedure will now have to be repeated for each combination of materials and settings. Having these figures available we can now go to the first step in the actual estimation.

First we look at the things we know. In most cases we know what kind of surface finish we have to end up with. We also have a pretty good idea what kind of wear can be allowed on the electrodes. Knowing that, the proper settings with known metal removal rate can be determined.

The next step is to calculate the amount of material to be removed. Do not forget to deduct flush holes in electrodes or workpiece from the total amount of material.

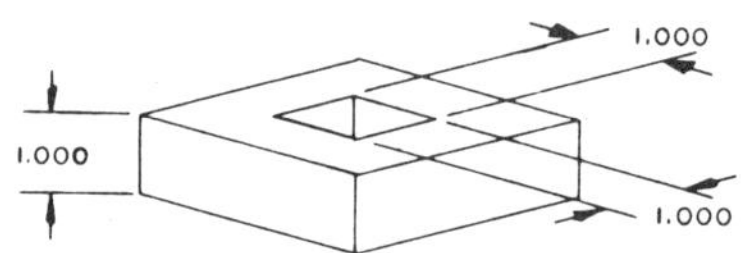

ELECTRODE CALCULATION

Establish surface finish of finished hole. Deduct the total overcut, related to the finish from the final dimension of the finished hole. That gives you the size of the finishing electrode.

Finished hole=	1.0000
Finish 31=Total overcut	– .0017
Finishing electrode	.9983

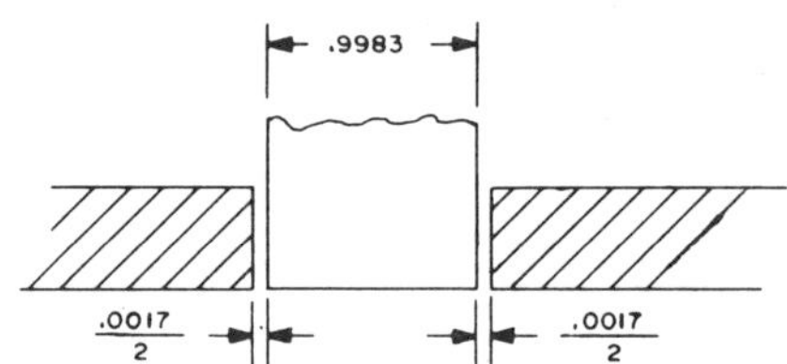

As a rule of thumb the difference in size between the roughing and finishing electrode can be established by mul-

METAL REMOVAL RATE IN RELATION TO A GIVEN ELECTRODE SIZE AND A .100 DEEP CUT

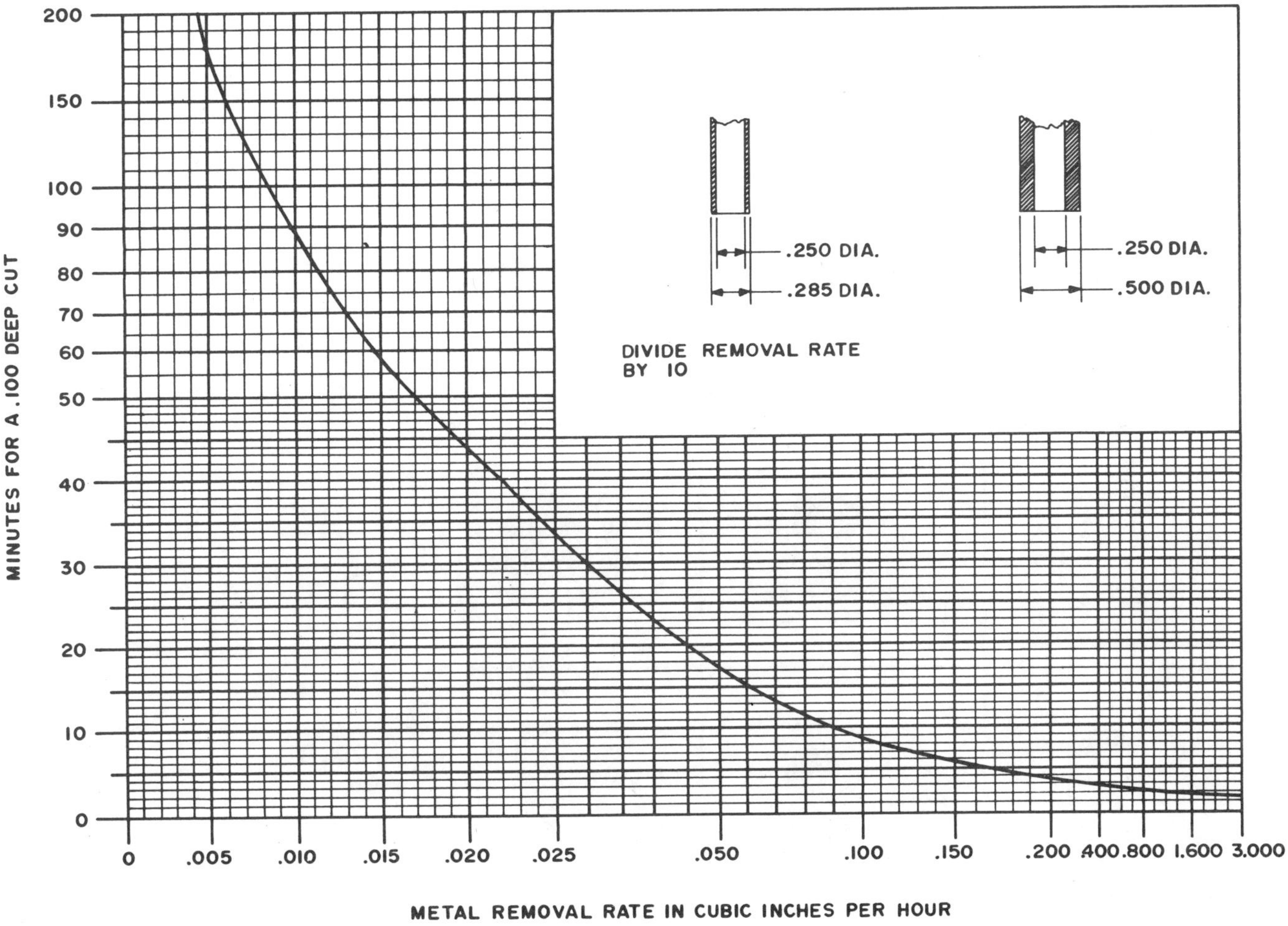

MINUTES FOR A .100 DEEP CUT

200
150
100
90
80
70
60
50
40
30
20
10
0

.250 DIA.
.285 DIA.

.250 DIA.
.500 DIA.

DIVIDE REMOVAL RATE BY 10

0 .005 .010 .015 .020 .025 .050 .100 .150 .200 .400 .800 1.600 3.000

METAL REMOVAL RATE IN CUBIC INCHES PER HOUR

tiplying the total roughing overcut by 2 or 3 depending on the safety margin you want to work with in relation to misalignment of electrodes and heat effect of roughing electrode.

Roughing overcut total	.0033
Finishing electrode	.9983
Roughing electrode	.9983=(3x.0033)=.9884

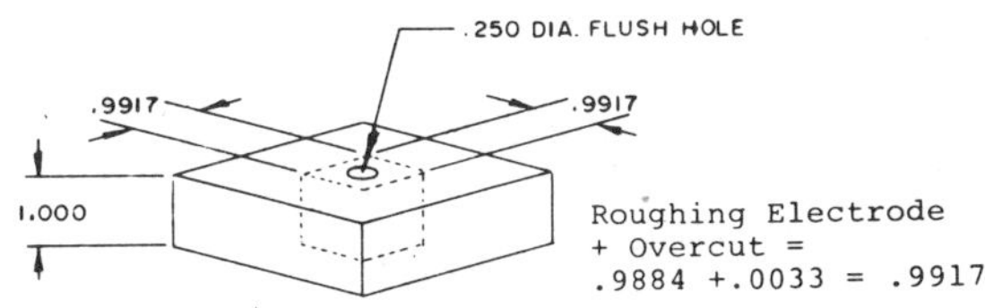

Material to be removed at the roughing cut:
(.9917″ x .9917″ x 1.000″)= $\pi/4$ x .250^2 x 1.000=
.9834−.049=.9345 cubic inches
Metal removal rate for the chosen roughing setting taken from your test cut is .620 cubic inches.

Material to remove	.9345 cubic inches
Material removal rate	.620 cubic inch/H
Cutting time	.9345÷.620=1 hour, 30 min.

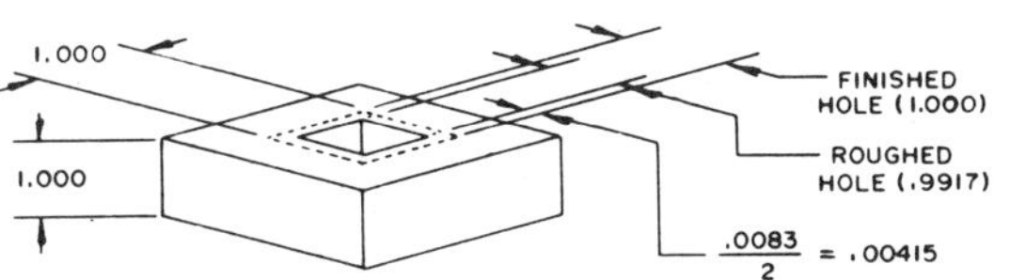

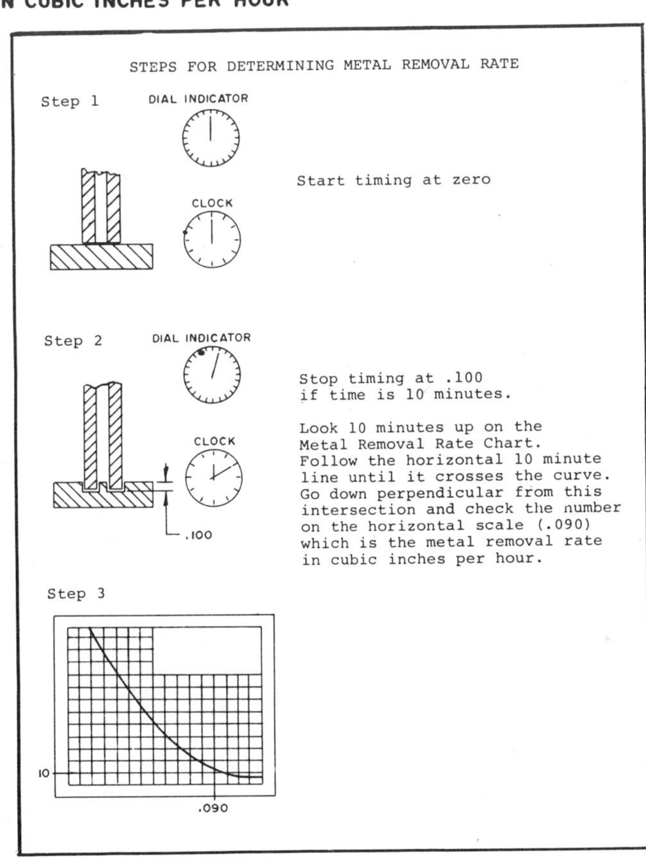

FLUSH EVALUATION CHART

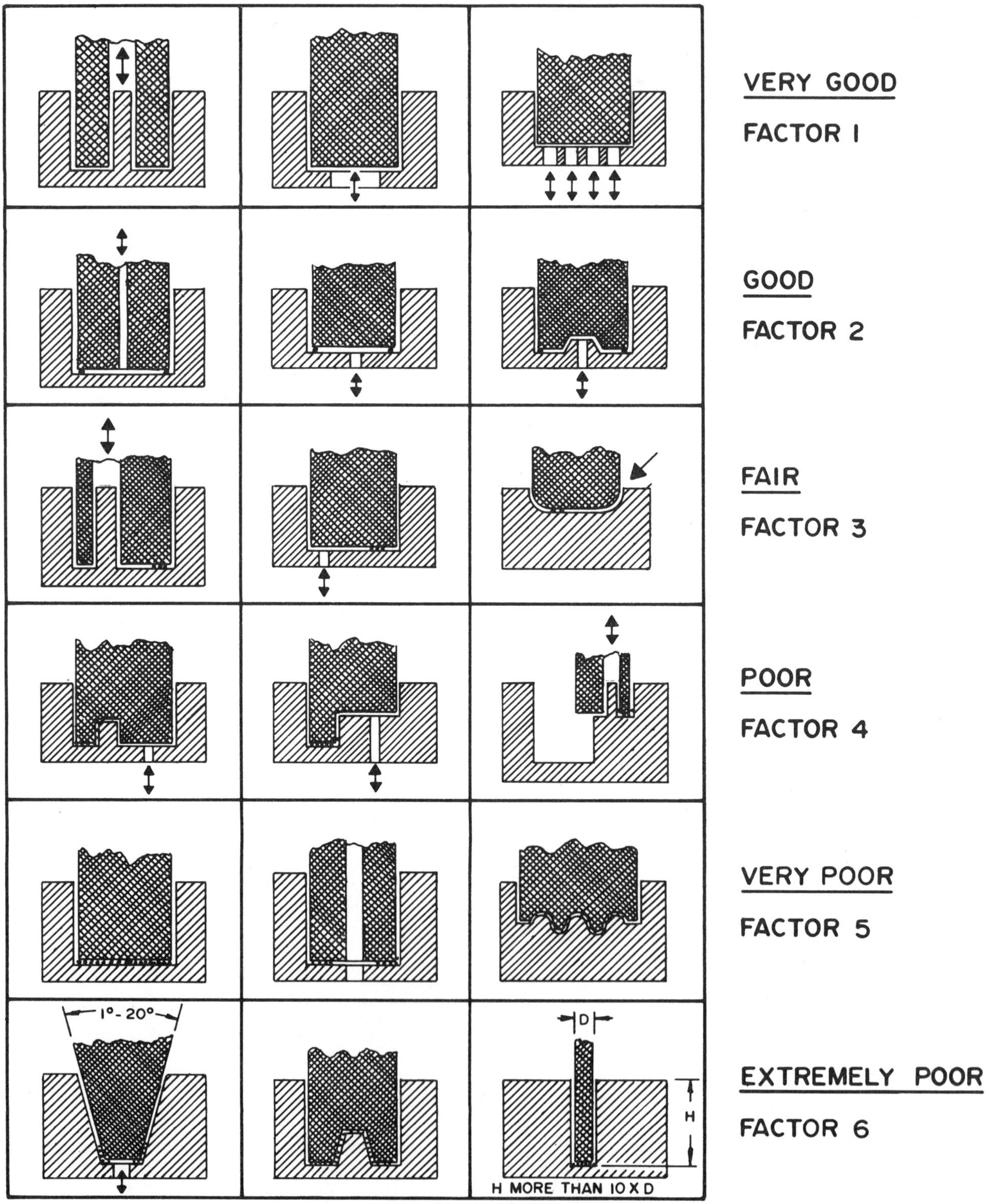

The easiest way to figure the material to be removed by the finishing cut is to figure the difference per side between the finished hole and the roughed hole and multiply this figure by the depth and length of the circumference of the hole and forget about the little inaccuracies in the corners.

Finished hole= 1.0000
Roughed hole= .9917

Total difference .0083
Per side .00415
Circumference 1+1+1+1=4"
Depth of cut 1"
Material to be removed: .00415 x 1 x 4=.0166 cubic inches

Metal removal rate for the chosen finishing setting, taken from your test cut was for instance .060 cubic inches/H.

Material to be removed .0166 cubic inches
Material removal rate .060 cubic inches/Hour
Cutting time .0166÷.060=.276 Hour=16 Min

We now have the theoretical cutting time for the roughing and finishing operation.

This is the moment where you have to decide what kind of flushing mode you will be using. The flush evaluation chart will give you the different possibilities. At the right side is an evaluation of the quality of the different flush modes. To each evaluation is connected a factor.

The theoretical cutting time for the roughing and finishing operation should be multiplied by this factor. In most cases from flush factor 3-6 you will have to use an interruption or pulsation of the cut to help the flushing. If your controls will be set so you cut for one second and interrupt for another second you are only cutting 50% of the time and your total cutting time will be in relation to that or in this case twice as long; so you multiply the "flush" modified figure by two. The final figure that you now have should be pretty close to the actual time.

As you can see the flushing plays a very large role in the total time. By putting a little more effort into improving the flushing, you can save a lot of cutting time by placing extra holes in the piece. But even the best estimate will be way off if you do not have an accurate tooling system to insure a perfect alignment of the electrodes. The two charts covering the influence of misalignment shows clearly, that if you want to finish at a fine surface finish, your project is doomed from the beginning if you work with a machine not supplied with the tooling needed to replace an electrode within "tenth". To try without is like going into a shooting gallery blind folded.

The figures shown on the two tables are very conservative and do not take any flushing difficulties into consideration. The first chart shows a cylindrical finishing electrode tilted 1.2 and 3 thousandths in relation to the previous electrode that is repeated for three different surface finishes and for six different diameters. The second chart shows the same for a flat electrode at different width.

To recap, it cannot be stressed enough, that to come up with an accurate time estimate it is of the utmost importance that you get the best possible flushing and an extremely accurate tooling system. Without it, you have only got half a machine.

(This article is based on a technical paper prepared by Mr. Houman for the Society of Manufacturing Engineers. Interestingly, the Eltee Pulsitron EDM equipment used in developing the time estimating procedure has, as standard equipment, several features that lend themselves to the concept including a depth finder that holds the electrode at a constant .0001" above the work during set-up. Ed.)

THE TIME INFLUENCE OF MISALIGNMENT OF CIRCULAR FLAT ELECTRODES AT THE FINISHING OPERATION.

SURFACE FINISH	X TILT	DIA. .5	DIA. .75	DIA. 1.00	DIA. 1.50	DIA. 2.00	DIA. 3.00
VDI 12 RMS 9	.001	20	40	1 HR. 10	2 HR. 30	4 HR. 20	10 HR
	.002	40	1 HR. 20	2 HR. 20	5 HR.	8 HR. 40	20 HR.
	.003	1 HR.	1 HR. 20	3 HR. 30	7 HR. 30	13 HR.	30 HR.
VDI 16 RMS 16	.001	5	10	20	40	1 HR.	2 HR. 20
	.002	10	20	40	1 HR. 20	2 HR.	4 HR. 40
	.003	15	30	1 HR.	2 HR.	3 HR.	7 HR.
VDI 22 RMS 22	.001	2	4	10	15	30	40
	.002	4	8	20	30	1 HR.	1 HR. 20
	.003	6	12	30	45	1 HR. 30	3 HR.

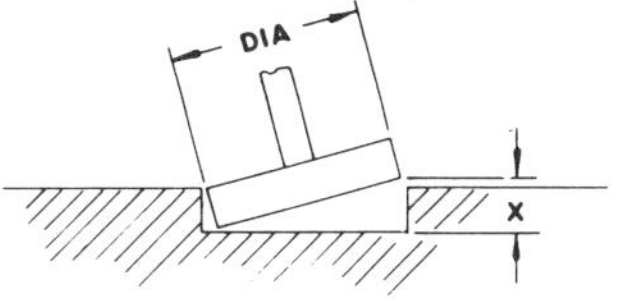

THE TIME INFLUENCE OF A ONE SIDED MISALIGNMENT OF A STRAIGHT FLAT ELECTRODE AT THE FINISHING OPERATION.

SURFACE FINISH	X TILT	Y .5	Y .75	Y 1.00	Y 1.50	Y 2.00	Y 3.00
VDI 12 RMS 9	.001	1 HR. 20	2 HR.	2 HR. 45	4 HR.	5 HR. 30	8 HR. 15
	.002	2 HR. 45	4 HR.	5 HR. 30	8 HR. 45	11 HR.	16 HR. 30
	.003	4 HR.	5 HR. 25	8 HR. 15	12 HR. 15	16 HR. 30	24 HR. 45
VDI 16 RMS 16	.001	20	30	40	1 HR.	1 HR. 20	2 HR.
	.002	40	1 HR	1 HR. 20	2 HR.	2 HR. 40	4 HR.
	.003	1 HR.	1 HR. 30	3 HR.	3 HR.	4 HR.	6 HR.
VDI 22 RMS 22	.001	8	12	16	24	32	48
	.002	16	24	32	48	1 HR.	1 HR. 30
	.003	24	36	48	72	1 HR. 30	2 HR. 20

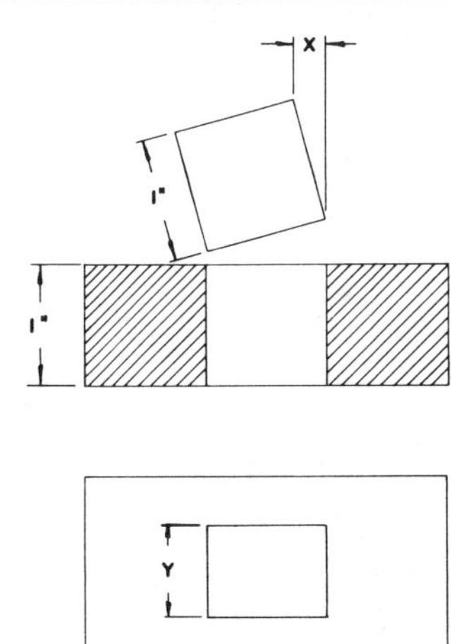

Reprinted from Modern Machine Shop

Optical Form-Grinding EDM Electrodes

By W. J. Knuff

The process of shaping an electrode may be 80 percent of an EDM job cost. Optical form grinding effectively tackles the electrode-shaping problem.

Electrical discharge machining (EDM) has emerged from an innovation to a highly practical and profitable process for producing intricate tooling. The process, in. essence, is the emission of a charge of electrons from a tool (electrode) which strikes a workpiece, causing an immediate rise in temperature high enough to melt or erode a small segment of workpiece material. The detritus is flushed away by the dielectric fluid in which both tool and workpiece are emerged.

In a controlled manner, the spark is pulsed from 160 to 200,000 cycles per second with the result that the eroded area assumes the opposite shape of the tool. There are many factors to consider in making EDM a productive reality. They include power supply, machine rigidity, feeding mechanisms, dielectrc flow and others. Two critical factors are electrode material and the shaping of that material to produce a geometrically accurate tool.

Electrode Material

The electrode must conduct electricity, and this permits use of a broad choice of materials. The most practical materials being used include brass, copper, copper-tungsten, silver-tungsten, graphite and copper-graphite. Each material has individual attributes related to cost, wear ratio, strength and detail reproduction. The most popular electrode material is graphite, with copper-tungsten and copper-graphite also finding widespread acceptance.

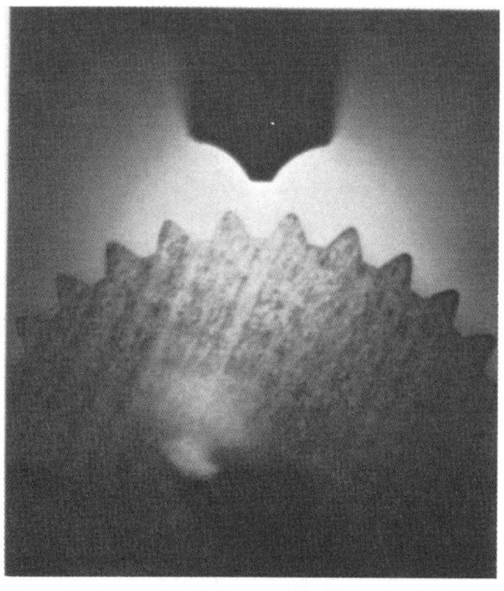

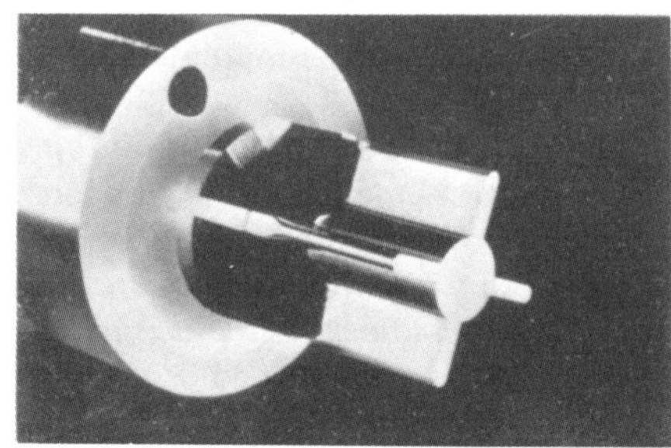

Graphite EDM electrodes that were ground on an optical form grinder. The diameter of the small electrode is 0.090 inch and the length is 4½ inches. The accumulated error is 0.0003 inch. Notice the formed grinding wheel used to grind the electrode.

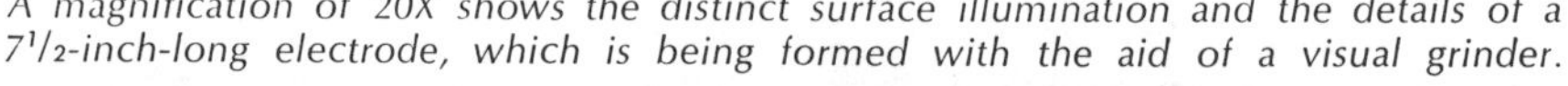

A magnification of 20X shows the distinct surface illumination and the details of a 7½-inch-long electrode, which is being formed with the aid of a visual grinder.

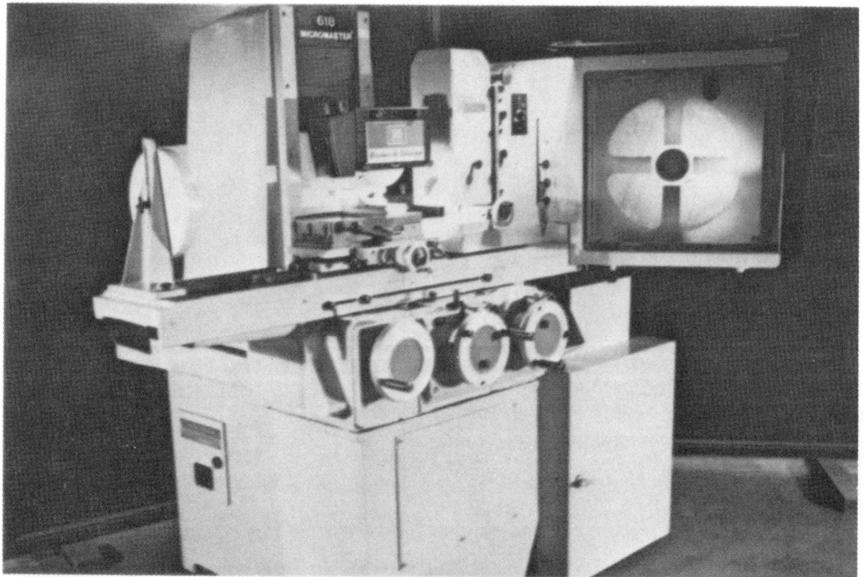

Shaping the electrode material into an intricate-shaped, precision tool is an important part of the EDM process. Far too often, economical and efficient electrode fabrication has been a low-priority item. Producing an electrode can represent 80 percent of the cost of the overall EDM process.

Many methods have been used to shape and form electrode materials; one method involves the use of optical form grinding (OFG).

Optical Form Grinding

An OFG machine utilizes an optical system to project an image of a workpiece onto a screen at a selected magnification. A line drawing of the final electrode form at the selected magnification is also affixed to the screen.

By observing the viewing screen, the operator can manipulate the grinding wheel to follow the chart on the screen. Thus, a very accurate form, which will not be affected by grinding wheel wear or mechanical changes, is produced on the electrode material.

An optical form grinding machine provides productivity, cost control and quality assurance paced by the capabilities of a specialized machine. Yet, all the personal skill of an individual can be utilized, since optical form grinding amplifies the operator's feel and vision. The result is a very high degree of control over the grinding process. The attributes of electrical discharge machining and optical form grinding lend themselves to great flexibility and innovation.

The Brown & Sharpe optical form grinder is basically a rugged, precise surface grinder that projects a magnified image of the workpiece onto an 18-square-inch screen in clear view of the operator. Magnifications of 5, 10, 20, 25, 50 and 100X can be selected, depending on the size of the workpiece and the accuracy required.

A magnetic chuck is clamped to the machine table. For some applications, the chuck may be attached to a subtable or cross slide along with a gage block and a 0.0001-inch-resolution dial indicator. This insures accurate positioning of the workpiece relative to the chart on the screen. The workpiece image is projected onto the screen in constant focus and in its true position (neither reversed nor inverted).

The hydraulic table can be operated automatically for linear grinding up to 12 inches, or manually for very fine control, which is particularly useful when grinding small, intricate forms. When using the reciprocating wheel slide, the table is stationary and the grinding wheel reciprocates along the linear axis of the workpiece. The reciprocating wheel slide, when equipped with a high-speed arrangement, is ideal for grinding a heel or shoulder on a punch or electrode.

Preparation of Charts

The prime advantage of using a chart to control form grinding is the high reduction ratio, which reduces layout error. In the preparation of charts, the usual drafting machine is not accurate or stable enough to draw precision outlines. It is better to use a chart layout machine. Charts can be made of frosted glass or plastic sheet —either clear or frosted. Pencil, ink, or scribe methods are all used. Standard toolmakers' charts, and angle, grid and radius charts are available and will suffice for many applications.

When the form area is too large to be projected on the screen, the contour is subdivided so that each section does not exceed the range of the magnification and screen. The individual sections are superimposed on the chart and ground in sequence.

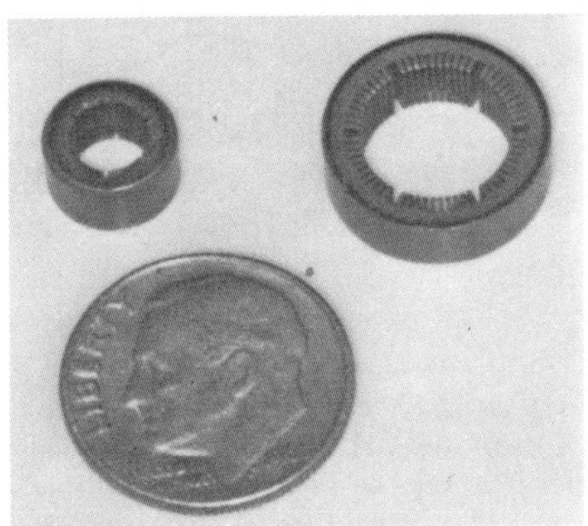

The internal slots of this unusual brass electrode are less than 0.003 inch wide.

A steel progressive die produced at Timms Spring Company. A visually ground graphite electrode was used to produce the die cavity and tungsten-copper electrodes were used to produce the punches.

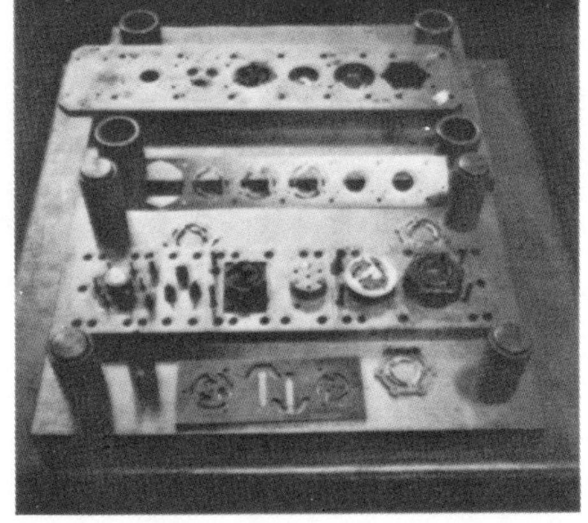

The small contact area between the grinding wheel and the electrode is conducive to grinding thin shapes such as this electrode, which is made of tungsten-copper.

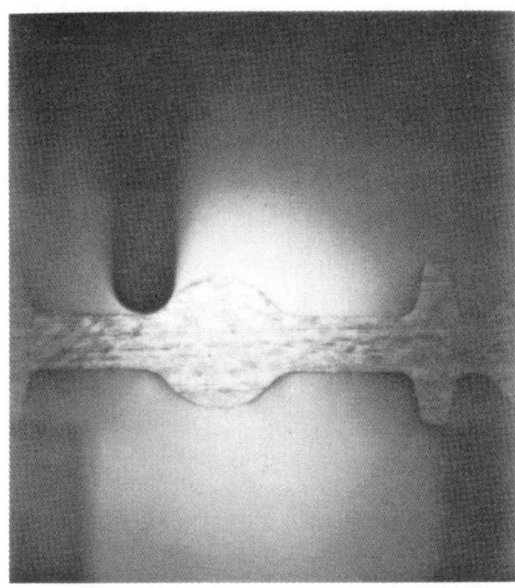

Tool and Die Use

Timms Spring Company of Elyria, Ohio, specialists in the engineering and manufacturing of quality spring products for over 60 years, have used a visual grinder to produce EDM electrodes for over 14 years. They produce all of their specialized tooling for vertical four-slide machines in their modern, well-equipped tool room. Much of their tooling is machined by using the EDM technique.

The tooling requirements at Timms Spring are quite demanding. Stampings are often produced in progressive dies at 300 or more parts per minute from stock only 0.002 to 0.006 inch thick. Barney Hostetler, tool room foreman, maintains that 0.0005 inch tolerance control is normal when using EDM electrodes to produce punch and die tooling.

Mr. Hostetler uses a single graphite electrode produced by OFG to EDM both the punch and die tooling members. A chart of matte-finished plastic is prepared for the OFG machine. A graphite electrode from three to six inches in length is prepared on the visual grinder, usually with a magnification of 20X, although this will vary according to the job. The electrode is then used for a rough layout on 1/4-inch-copper-tungsten plate, which is then rough-sawed to remove most of the metal from the opening. The plate is then EDM'd with the graphite electrode to bring it to its final close-tolerance configuration.

The finished copper-tungsten plate then becomes the electrode to EDM a heeled punch of hardened die steel. The graphite electrode (which has had very little wear, since most of the metal in the copper-tungsten plate was sawed away) is used to EDM the mating die opening in hardened die steel. By controlling the power settings on the EDM unit, the overcuts provide the necessary tolerance fits.

The design complexity, tolerance specifications, life requirements of the tooling, and other parameters surrounding the individual job determine the exact combination of EDM and OFG used. For example, optical form grinding may be used to generate an electrode, which will be used to form a die. The die in turn may become the electrode to EDM a mating punch. Or the punch itself may be generated by optical form grinding and tipped with an EDM electrode to form the die. In unusual instances, a punch and a split die may be formed by optical form grinding with EDM used to form the stripper. The variations are many and are based on the particulars of the job.

Powder Metal Tooling

Powder metal tooling is quite different from stamping dies and it has some unusual requirements that can be met with a combination of electrical discharge machining and visual grinding. For example, some of the requirements are:

- A very close fit between punches and die; usually between 0.0001 and 0.0005 inch.
- A high finish, often 2 to 5 rms.
- A very high index accuracy, particularly in gear dies.
- Long lengths for punches, dies and core rods.
- Punches with strong support heels, which are awkward to machine and inspect, to withstand the high pressures involved in compressing powder metal.

Powder metal dies of tool steel or tungsten carbide can vary from two to eight inches through the opening. As a result, more than one electrode may be required to generate a punch or die. These may be a rough, re-rough, finish and finish-lap electrode, each a slightly different size of the same configuration. The OFG chart will show each different size with a specific and uniform size and contour control.

Tipped Punches

A hardened steel punch is often ground on an optical form grinder and tipped with a graphite electrode to EDM a die. The process starts with cementing the electrode material onto the front of a hardened punch blank, using an adhesive that will conduct electricity. Eastman 910 adhesive can be used. The punch and electrode can be form-ground at the same time.

Since the punch and electrode are the same size, a controlled over-cut will produce the clearance required between punch and die opening. If a smaller electrode is required, acid etching of the electrode area may be considered or the electrode may be reground.

Tipped punches can be used independently in an EDM machine to produce an individual die opening. Multiple punches can be located in the punch holder of a complete die set and used in the EDM machine to produce the openings in the die holder. This results in excellent positioning and uniform clearance of the punches relative to the die openings, since all components are already in the die set. After EDM'ing, the electrode tips are removed from the punches and the die is ready for use.

When using tipped punches to EDM the mating openings in the die

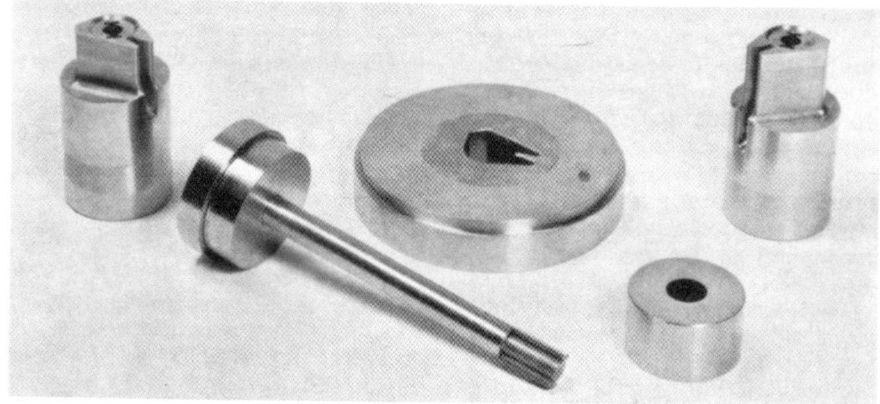

Typical tooling utilizing carbide die and powder metal. The die and the internal core of the punches were EDM'd with OFG-shaped electrodes. The outside contour of the punch and the carbide tip on the core rod were generated by optical form grinding.

A stamping die with visually ground punches and split die. The stripper and bottom die shoe were both finished by making use of the electrical discharge machining process.

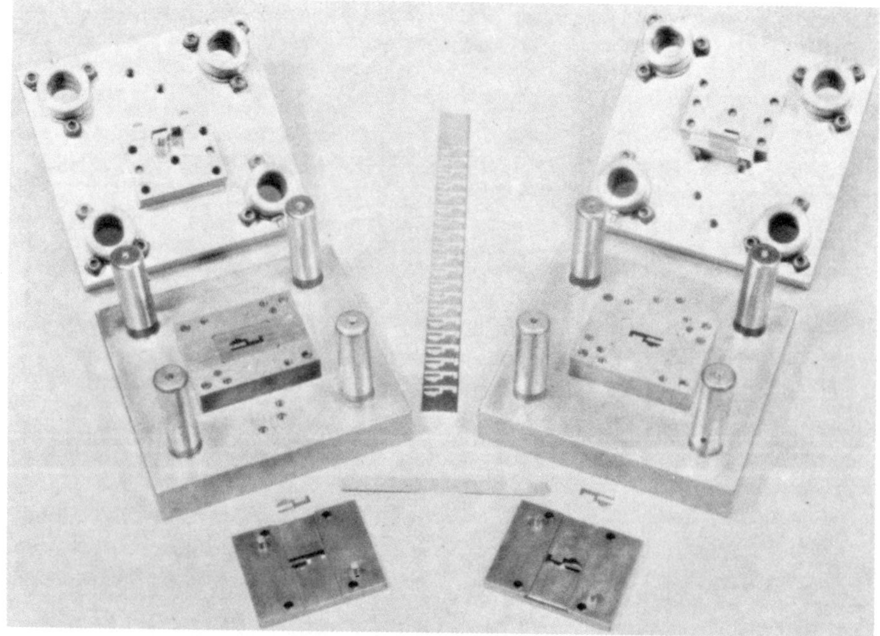

holder, there must be complete electrical insulation between the upper and lower die shoes. Otherwise, there would be a short circuit in the EDM unit. Electrical current control is achieved at Timms Spring Company by employing Lemcoloy insulated bushings on the die posts.

Formed Wheels

While OFG is most often used with a flat or radius-face grinding wheel, a formed wheel may also be used. A scribed chart is used to verify the final form. There are many ways to dress a specific form on a wheel, including crush form, single-point diamond, diamond block and pantograph. The dressed wheel can be checked by projecting the grinding wheel on the screen and matching it to the chart.

Electrode materials, particularly graphite, can be ground very easily. After the form and position are determined, the process is very fast with relatively little wear of the grinding wheel. **mms**

618 VISUAL GRIND — BRIEF SPECIFICATIONS

BS CAPACITY	ENGLISH MACHINE	METRIC MACHINE
Width and length of Work Ground	6 x 14	150 x 355
Spindle 618 ORIFLEX Drive (2860 RPM)	1 HP	.75 kW
Spindle 618 DIRECT Drive (3600 RPM)	1½ HP	1.1 kW
Spindle 818 DIRECT Drive (3600 RPM)	2 HP	1.5 kW
Feeds Longitudinal Feet/Metres per min.	5-50 (HYD)	15-150
Longitudinal Travel	16	405
Cross Feed Travel	(618/818) 7/9½	(618/818) 50/70
Vertical Travel	15¾	400

Reprinted from: Tooling and Production, May 1970

EDM speeds Phoebus nozzle machining

Are you cutting grooves, scallops, slots and tricky holes in materials like Hastelloy and Inconel with conventional machining methods—and losing your shirt? This manufacturer found the answer in an EDM setup engineered to perform several operations at a time.

AN UNUSUAL EDM system recently enabled Tracer Corp., Santa Monica, Cal., to process a large complex venturi-shaped nozzle in a third of the time required for conventional machining. The 7000-lb Phoebus nozzle is part of a giant nuclear engine made by Aerojet-General Corp. for future space trips.

Specifications

The 91″-long nozzle has a 60″-dia forward end, tapers to an 18″-dia throat and flares out to a 46″-dia exit. It is made by electron-beam welding three Hastelloy-X forgings together, and machining the welded nozzle to a smooth surface.

Specifications require cutting 320 grooves, each 0.040″ wide ±0.001″, axially around the inside of the nozzle. Groove depth must be 0.100″, with the radius of the groove bottom held to 0.010″ max. Each groove requires 8 scallops—a total of 2560 scallops for all the grooves in the nozzle.

At the large end of the nozzle, 320 holes, ½″ dia, must be put through 1″ of solid Hastelloy between the grooves.

At the small end of the nozzle, 320 slots, 0.260″ x 1.14″, must be made through 2½″ of solid stock between the grooves. And 144 holes, ⅛″ dia, must be run through 4″ of stock at an angle, to intersect 144 blind holes.

Old method

When the operator used a conventional tracer lathe, he had no way to continue groove cutting through the throat of the venturi. He got around this problem by cutting grooves as far into the throat as possible, turning the nozzle around, resetting for alignment, and then finishing the other end of each groove cut.

He had to cut the 2560 scallops one at a time, and then deburr them. While the 320 half-inch holes posed no great problem, the 320 slotted ones did.

And if the customer had changed the basic-material specs to tungsten or Inconel instead of Hastelloy-X, the engineers would have had to design different tooling and set up new cutting practices.

New method

With the new EDM system, the operator installs the Phoebus nozzle in a fixture that indexes it accurately 320 times through 360 degrees, **Figure 1.**

A grooving electrode, running the length of the venturi, hangs in the nozzle, **Figure 2.** The aluminum tube that holds it receives dielectric fluid that is pumped from a well beneath the fixture. A hydraulic head from an Easco-Sparcatron 162 EDM machine, placed over the top center of the indexing fixture, controls electrode movement accurately.

At the large end of the nozzle, an overhead rotary EDM head independently machines the ½″ holes. At the small end, a third overhead EDM head cuts the rectangular slots, **Figure 3.**

Each head has its own power unit, the center and the rotary heads having a Sparcatron SPF 25 and the rectangular-slot head a rotary impulse generator. And with the two outer heads supplied by pumping dielectric fluid into dams around the electrodes, the machine can now EDM grooves, holes and slots at the same time.

Machining the axial grooves requires two cuts, one for roughing close to the required depth, and one for finishing to provide the sharp corners, **Figure 2.** The operator removes the electrode holder after each roughing cut for template-guided redressing, which takes only minutes. He then puts it back in perfect alignment with the two large columns (at the ends of the venturi) that are part of the fixture and have positive positioning plates, **Figure 4.** After the finishing cut, the venturi indexes and the electrode makes the next roughing cut.

Scallop electrodes, adapted to the grooving-electrode holder, enable cutting the eight scallops in each groove at one time. Thus 2560 separate cuts are now reduced to 320, a job that takes about 40 hours—and no deburring is required. This was about the time required for deburring alone in the conventional machining operation.

Machining the 144 ⅛″-dia angular holes completely through the 4″-thick shell is easily accomplished, without additional tooling, because the engineers altered the angle of the rotating-spindle EDM head.

This multiple-head EDM approach enabled Tracer Corp. to produce three Phoebus nozzles in the time taken to make one by the old method. ■

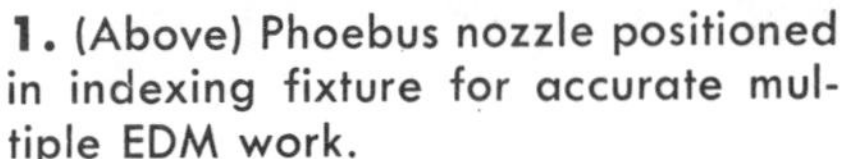

1. (Above) Phoebus nozzle positioned in indexing fixture for accurate multiple EDM work.

2. (Left) To EDM a groove, the dielectric flows into the work cavity from an aluminum tube that holds the grooving electrode.

3. (Above) Three independent EDM heads—shown at top left, center, and right—make it possible to perform three operations at one time.

4. (Left) Electrode holder in working position. The positive-positioning plate of the electrode-advancing column appears at lower right. Half-inch-dia holes are EDM'd between grooves at the outer edge of the nozzle.

Reprinted from a translation published by Welding Production, February 1976

Arc Stability In Electric Discharge Machining

By I. V. Ryabov
The V. Ya Chubar Institute of Mechanical Engineering

The characteristic features of machining with a disc-shaped tool electrode, with the supply of a 3% solution of self-emulsifying oil in water to the machining zone (1), are the use of a d.c. power source, rapid movement of the workpiece (0.5-2m/sec) relative to the tool, and low working voltage (about 20V). The rotary speed of the tool disc is 30-50m/sec.

This process can be employed for machining intractable materials at a rate of 2-3cm^3/sec, with roughness of the machined surface R_z = 40-160μm, and depth of the layer with altered structure not more than 0.1mm (2,3). The allowance is removed by local burning-off by low-voltage arc discharge. The tool disc is the cathode, and the workpiece is the anode.

To select optimum machining conditions, it is necessary to know the technological characteristics of the low-voltage arc discharge in conditions of relative movement of the electrodes. Arc stability is an important technological factor, and it has an influence on the quality of the machined surface.

The present article gives the results of studies of the stability of arc burning in conditions of electric discharge machining.

Special equipment was used for the experiments. This equipment permits movement of the specimen at a speed of up to 10m/sec. The rotary speed of the cast-iron tool disc is 43.5m/sec. A direct-current power source was used, connected in a three-phase bridge circuit using semiconductors (VK-200) with smooth adjustment of voltage from 15 to 36V. The specimens were made in the form of a sector of a disc with a length along the circular arc of 40-50mm and a width of 10-20mm. For arc striking, a 'starter' with a section of 0.5 x 0.5mm and a height of 0.1mm was centre-punched on the specimen. When the specimen moved along a circular arc relative to the tool disc with a 0.05mm gap, the starter came into contact with the working surface of the disc, and

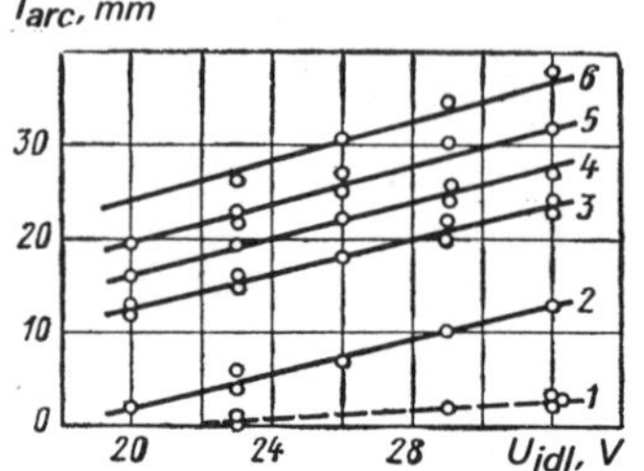

Fig.2. Arc length at extinction (l_{arc}) as a function of idling voltage (U_{idl}): 1, arc burning in liquid; 2-6, arc burning in air, with anode speeds of 0.5, 1, 2, 3.5 and 7m/sec respectively.

this facilitated arc striking. We investigated the stability of arc burning in the gap between the specimen and the working disc, both in air and with cooling of the arc burning zone with fluid. As the specimen moved away from the disc, the arc was able to burn between the end of the specimen and the working surface of the disc; stretching of the arc column then occurred.

Arc length at the moment of extinction can be taken as a criterion of its stability: the longer it is, the more stable is the arc discharge. During machining, however, high stability may lead to excessive fusion and impairment of the quality of the machined surface.

During movement of the specimen relative to the tool disc, at the moment when the starter of the specimen comes into contact with the disc, short-circuiting occurs, and this leads to rapid heating and then to explosion of the starter and the formation of an arc discharge. Figure 1 shows oscillograms of the process of arc burning in air and in liquid. The idling voltage of the power source is 26V, and the specimen speed is 9m/sec. With an arc burning in air, the maximum current is 1220A, and the process time 8.9msec, but in liquid the current did not exceed 900A, and the process time decreased to 2.1msec. The initial period of arc formation is characterised for both cases by the presence of metallic contact between the electrodes, which lasts

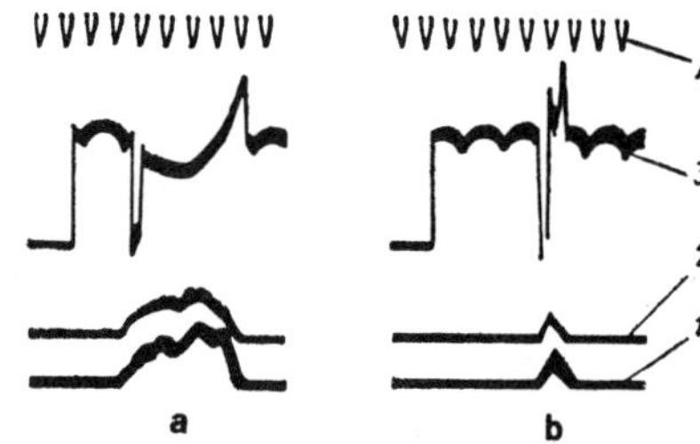

Fig.1. Oscillograms of arc burning in air (a), and in liquid (b): 1 — arc power; 2 — current; 3— voltage; 4 — time marks with frequency of 500Hz.

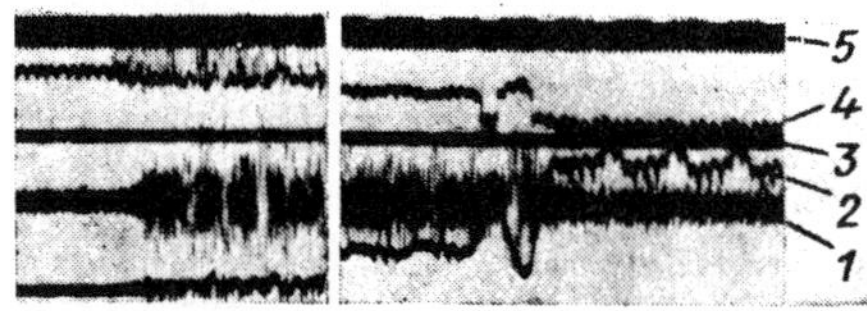

Fig.3. Oscillograms of the process of electric discharge machining with continuous variation of productivity: 1 — recording of arc noise; 2 — current; 3 3 — zero voltage; 4 — working voltage; 5 — time marks with frequency of 500Hz.

for 0.7-1.0msec. At the time of short-circuiting, the voltage drops to 2-3V. After explosion of the starter, arc striking occurs, and the voltage is established at 19-23V for arc burning in liquid, and at 16-17V in air. The voltage depends on the properties of the anode material. Thus, in the machining of tungsten and molybdenum the voltage is 19V, for nickel base alloys it is 20V, for copper 21V, and for aluminium 23V.

Analysis of the oscillograms of the arc burning process has led to the establishment of the dependence of arc length l_{arc}, in mm, at the moment of extinction on the machining conditions: $l_{arc} = 7.9 + 6.1 I$, where I is the arc current, kA.

The arc length at extinction increases with increase of voltage and speed of movement of the anode.

The shortest arc length at extinction is observed when liquid is supplied to the arc burning zone. In this case, when the voltage is less than 22V the arc length at extinction does not exceed 0.1mm. The presence of liquid further reduces the arc length (Fig.2).

The stability factor S can be employed for assessing arc stability:

$$S = \frac{l_{arc} - \delta}{\delta},$$

where δ is the working gap between the electrodes.

In semi-finishing electric discharge machining at the optimum working voltage of 19-23V, a gap of 0.02-0.03mm is established between the electrodes. In these conditions, the value of the arc stability factor is fairly low, and the arc can only burn in a stable manner at gaps less than 0.05mm, as can be seen from the oscillogram shown in Fig.3.

Machining productivity varied from zero to a value at which all the power of the supplying source was utilised, and the voltage then dropped from 28V to a limiting value of 20V, after which short-circuiting occurred.

A gap proportional to the working voltage was established between the workpiece and the tool.

Initially, when the voltage is fairly high and the machining process is unstable, there is discontinuity of voltage and current, and arc extinction is accompanied by considerable noise.

As the working voltage is reduced, the electrode gap decreases, and the machining process stabilises. The voltage fluctuations decrease, arc extinctions cease, and there is less noise. When the voltage decreases to 20V, with the establishment of an electrode gap in the range 0.02-0.03mm, the machining process is stable; any further reduction of voltage leads to short-circuiting.

Electric discharge machining with a minimum electrode gap ensures the lowest possible specific energy consumption, and the quality of the machined surface meets the requirements for semi-finishing operations.

CONCLUSIONS

1. A characteristic feature of semi-finishing electric discharge machining in a liquid medium is localisation of the arc discharge in the machining zone because of a reduction of arc stability.
2. At a reduced working voltage of 19-23V, and movement of the workpiece at a speed of 2m/sec in a liquid medium, the arc discharge is stable at an electrode gap of $\leqslant$ 0.1mm.

REFERENCES

1. BORISOV, B.Ya. and RYABOV, I.V. The electric discharge machining of refractory metals with direct current, with supply of liquid. In symposium: Electrophysical and electrochemical machining, No. 6, Moscow, NIIMASH, 1970.
2. RYABOV, I.V. The influence of process variables on surface quality in electric discharge machining. In symposium: Machining of metals, No. 1, Moscow, TsNIITEITrakorosel'khozmash, 1971.
3. BORISOV, B.Ya. and RYABOV, I.V. Investigation of the characteristics of a d.c. arc discharge in electric discharge machining with the supply of liquid. In symposium: Electrophysical and electrochemical machining, No. 4, Moscow, NIIMASH, 1970.

Reprinted from: Tooling and Production, January 1973

Cap and gown are worn in this assembly area.

EDM FOR EMISSION CONTROL

It can cost us all a lot of money to meet the new standards for clean air. Fortunately, efficient manufacturing methods can help soften the blow. As an example, here's a story of how electrical discharge machining combines with mass production in a clean room. The result is a special valve to help control nitrogen oxide emissions in all 1973 passenger cars and trucks from Chrysler Corp.

1. At the EDM station, outside the clean room, operator loads special turntable fixture while the machine automatically unloads. Kerosine helps cool the workpiece.

NO SMOKING is but one of many watchwords remembered by the elite group of workers electing to assemble the orifice spark advance control (OSAC) valve.

The OSAC valve must be absolutely clean when it goes together, so there are tight quality control procedures and certain restrictions on the personnel. For instance, there is a dress code: no excessively long hair—and cap and gown must be worn at all times. No eating or drinking in the clean room. In fact, the workers don't even breathe too hard; if they have a cold or flu, and still come to work, they must be reassigned to conventional tasks outside the clean room.

The reason? Inside the special area, the air is filtered to meet the standards of a Class 100 clean-air room. To meet this specification, there can be no more than 100 particles per cubic foot any larger than 0.5 micron, and

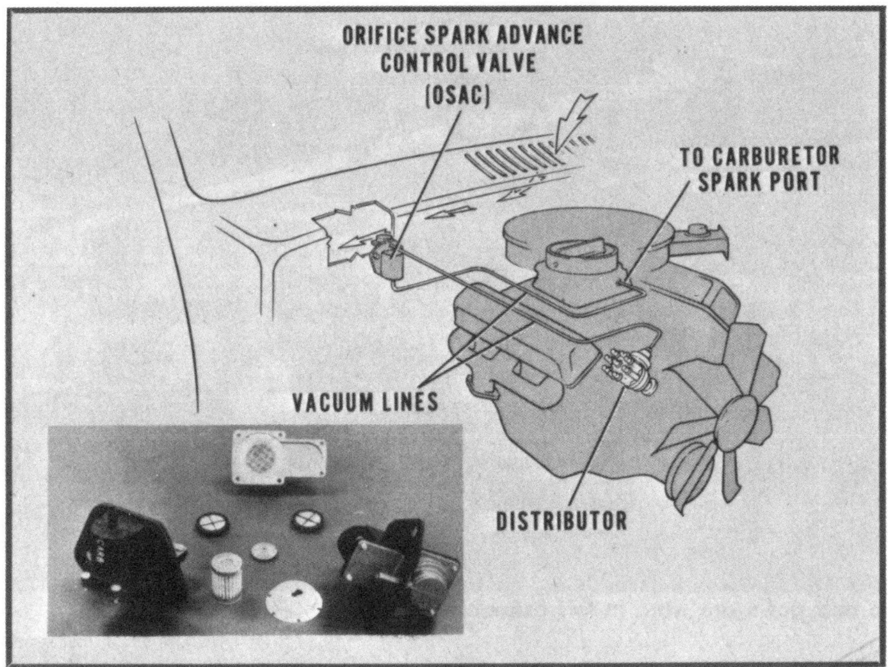

2. Where OSAC fits the system. As engine speed increases, vacuum created by higher air flow in carburetor tries to pull air from the distributor to activate the spark-advance system. The new valve delays distributor-vacuum buildup by 17 sec to retard spark advance and decrease NO_x output while the engine develops increased torque. Internal bimetal valve bypasses OSAC valve when outside air temperature is below 58 F. Otherwise, car would be hard to start and difficult to operate in cold weather. Fortunately, NO_x emissions are minimal in cool-weather operation. Location of valve in heater plenum keeps it at temperature of outside environment. Photo shows completed valves and some internal components, including three filters that help each valve serve for over 50,000 miles.

3. Arrow points to microscopic hole in orifice cup.

only 1 particle can be as large as 4 microns.

Outside, it's much less pure, but probably considerably better than in the average factory, in which a cubic foot of "air" typically contains 3 million particles as large as 300 microns, and many millions of smaller particles. In checking all this out, operators in the clean room, viewing through 300-power microscopes, study particles the size of large bacteria.

According to Elmer Sivacek, general manager of the Introl Div., Ann Arbor, Mich., "The operators prefer to work in the clean room. They know they are taking part in a special project, and they are proud of the job they are doing." That job is indeed quite a task. In spite of all the filtering, laminar air-flow devices, degreasing and ultrasonic cleaning of parts before they enter the clean room, and vacuum cleaning of parts immediately before assembly, some valve assemblies still get through with enough dust or dirt to clog up the small orifice. Thus, quality control is as much a part of the production operation as the actual assembly steps. Once the valves are assembled, built-in filters keep the maximum particle size down to 0.4 micron. Filters last at least for 50,000 miles of engine operation, say Chrysler officials.

EDM for a tiny hole

The OSAC valve provides a delayed-action spark advance in response to increasing vacuum, thereby delaying the burning process in the engine cylinders. To do this, the new antipollution device delays the vacuum buildup in the distributor by 17 sec. Heart of the system is a 0.003″ orifice or restricting hole burned through the face of a 1″-dia aluminum part called the orifice cup **(Figures 1, 2 and 3)**. The small hole delays the vacuum buildup at temperatures above 58 F.

"We could have used lasers to drill the holes," says Mr. Sivacek, "but our investigation revealed that this approach would cost too much. Ordinary twist drills would be too fragile and imprecise for this work, especially in quantity production. We considered chemical etching, but found it not completely reliable. And punching would have been difficult in the 0.030″-thick part. So it really appeared that EDM would be the only practical way to do the job."

The hole must be burned to a tolerance of ±0.000,06″. However, it need not be exactly round, and the critical parameter is actually a measure of air flow through the cup. A test setup stands next to the EDM station ready for periodic checks to see that air flow through the microscopic hole is on target. A measured quantity at specific pressure must flow through in 17 sec, ±1 sec. Any deviations require adjustment of the EDM unit.

Actual hole size depends on the lot being machined. The aluminum is first coined to a thickness of 0.0035″, ±0.0005″, from 0.020″ stock. Depending on the actual coined thickness and the hardness of the aluminum, the required hole size for the lot varies from 0.003″ to 0.0035″.

"The EDM is essentially a standard Raycon machine with tooling designed to meet our specifications," says Mr. Sivacek. "We chose wire electrodes and developed the indexing table. The wire is used up as we go, but its cost is relatively low."

The tungsten steel electrode wire is 0.0034″ dia, and accuracy of the wire size is important to EDM performance. However, further control is established by current delivered from the power supply, and cycle time ranging from 12 to 15 sec per part. Current is controlled for each of two wires processing two parts at once.

"We vary time and current using a standard Raycon control," says Sivacek, "but we had to develop rules and build up experience to get the special tooling working just right. Now the operation is consistent, and we would only have to check the machine once an hour. In practice we do it every half hour for extra assurance, determining machining performance by measuring drift away from the actual size (air flow time) desired. We know which way to change the current or time to correct any drift from the proper size hole."

4. The clean-air room has the standards of a hospital operating room. Room temperature of 68 F matches one of the valve test conditions.

High-speed hospital

After EDMing, the parts receive a thorough cleaning. In fact, many components of the OSAC valve are air blasted in a tumbler unit and then given a 2-min ultrasonic bath. Next, an operator wraps the parts in a plastic bag and puts them into a chamber leading into the hospital-like clean room **(Figure 4)**. The double doors on the passageway are interlocked so that both cannot be opened at the same time.

Inside, the operators unwrap the parts and begin the 4-hr, 10-step assembly and test process, making an average of 8000 valves each day in a single 8-hr shift. In one important step, they batch load orifice cup subassemblies into a glass incubator cabinet to bring them down to a test temperature of 48 F. Later, the clean room temperature of 68 F provides the atmosphere for a second test. Batch loading allows the flow of 8000 parts a day in spite of the long "soak" periods.

At a nominal temperature of 58 F (minimum 48 F), each OSAC valve must bypass itself to ensure good engine performance in cool weather. Thus, in the 48 F incubator test, a bimetal disc in each valve must move outward to expose a bypass hole, and it must move inward again at 68 F (maximum) to reactivate the vacuum delay.

Finally, checks for air leaks require an inspection and dump operation in which the vacuum flows through each valve in reverse **(Figure 5)**.

Completing the valve

Orifice subassemblies are heat-staked to a die-cast cover assembly that contains a small filter to screen air from the distributor. This disc filter keeps particles as small as 1/200 the diameter of a human hair from reaching the orifice hole. The flange subassembly has a barrel-shaped prefilter and another disc filter for screening air from the carburetor. Mating of the flange and cover occurs along a 40-ft ferris-wheel-type assembly line moving about 2 mph. Here each subassembly rides in a color-coded tray.

Valves are tested in lots, and one bad valve means a careful reinspection of the entire lot. Successful valves leave the clean room in plastic containers and go to a hot stamping machine to receive an imprint of the completion date and color-coded legends to assure proper hose connections.

The division people are proud of the precision work they do on a mass-production basis. It's challenging work, and the products range from speedometers, gages and instrument clusters to PVC valves and activated carbon canisters to absorb gasoline vapors from carburetors and gas tanks. Interestingly, only 60 percent of the plant output serves Chrysler Corp. products. The remaining 40 percent goes to other auto makers! ■

5. Leak tester automatically runs vacuum through each valve in reverse. Ferris-wheel conveyor system appears in foreground.

A Practical Application Of EDM To Manufacture Repeatable Burr Free Metering Orifices In Carburetor Components.

By Louis Mihalyfy
Manufacturing Process Engineer
Ford Motor Company

The advent of Government legislated emission control standards necessitated the search for new methods of producing more uniform, close tolerance and burr free carburetor metering orifices.

Conventional machining techniques did not yield repeatable orifices within a tight flow band required for critical metering purposes and the use of more exotic machining techniques was found to be costly and time consuming, prohibitive of high volume low cost production.

A preliminary evaluation of Electrical Discharge Machining (EDM) disclosed that it would produce orifices with less variation than conventional processes as well as any other exotic processes studied.

Ford initiated a program to develop EDM equipment suitable for production use, measuring equipment to evaluate EDM capability, and study a feedback system that would control the EDM operation.

The successful conclusion of this program proved that:

1. EDM machining could compete with current production processes and that orifice flow uniformity is at least three times better.

2. Burrs and associated labor costs of deburring are eliminated.

3. Continuous operation could be sustained for at least an eight hour shift without electrode change in a particular operation.

4. And finally, a closed loop control system could be designed to control and vary orifice size within narrow limits without electrode size change.

Subsequently, production equipment was installed and successfully used to EDM metering orifices in the following components:

Two rectangular metering slots in the aluminum die cast main body of the two barrel carburetor.

One bleed hole in a brass carburetor idle tube.

Two metering orifices in the zinc die cast booster venturi support, combining EDM and conventional machining in a rotary indexing machine with a computer controlled closed loop feedback system.

INTRODUCTION

Electrical Discharge Machining is best known for its ability to produce intricate dies, molds or prototype parts in small quantities even in exotic materials.

A study was initiated to apply EDM technology to machine small metering orifices in carburetor components. Conventional machining techniques - drilling, reaming, deburring - did not yield repeatable orifices within a tight flow band needed for critical metering purposes. The use of more exotic machining techniques proved to be costly and time consuming, and as a result was prohibitive for Ford's high volume production requirements.

Preliminary experiments produced EDM holes with less variation than the conventional process or the other processes studied. Although similar experiments were conducted as early as 1959, the lack of high speed production equipment and experience in machining small holes prevented the use of EDM for this application at that time.

The technological advances made in recent years, such as solid state power supplies, improved servo controls and better electrode materials, now made this process appear economically feasible.

Based on these events, a full development program was initiated to achieve the following objectives:

1. Develop EDM equipment suitable for production use that would produce more repeatable, burr free holes competitive in cost with the existing process.

2. Develop measuring equipment with the accuracy and repeatability required to evaluate EDM capability for this program and for future production requirements.

3. Develop a feedback system that would control the EDM equipment based on the measuring equipment readings. This system would be used in areas where close dimensional control is required.

DISCUSSION.

It was obvious from the beginning, that machining parameters recommended in EDM manufacturers literature were inadequate for the special purpose for which this process was intended. The main application of EDM was, and still is in the field of controlled metal removal, mostly in die making. Information was scarce in the area of accurate small hole machining and almost nonexistent for machining composite materials like zinc and aluminum die castings.

The task was then to construct an experiment using statistically valid techniques to optimize the process parameters for evaluating any make and model EDM equipment. The primary objectives of this program were:

- Determine the effect of EDM equipment setting variables on air flow repeatability of orifices. This parameter was chosen because it was felt that related to carburetor performance, the air flow characteristics of a given orifice more fully define the function of the orifice than a mechanical size measurement.

- Determine the equipment settings that produce the best flow uniformity/machining time combination.

- Select the best electrode material for the intended application.

- Determine the electrode wear per machining cycle in the acceptable operating range.

Preliminary search disclosed that available EDM equipment did not fulfill certain requirements of this special application, mainly in the area of holding small diameter electrodes, electrode wear compensation and proper flushing. Consequently, a special machining head was designed and built (Patent Pending). The head incorporated automatic electrode wear compensation, electrode rotation and through-the-electrode flushing. This completely self contained unit with its own solid state logic control circuitry, dielectric and hydraulic supply, enabled us to evaluate various EDM power supplies under the same operating conditions.

Simultaneously with the EDM development program, a computer controlled air flow measuring equipment was developed, capable of measuring changes in air flow caused by orifice diameter changes as small as .0005" (0.0127 mm). A hot film anemometer sensor was used and elaborate precautions taken to minimize or account for variables affecting flow measuring accuracy.

In the first part of the test program, the setting changes in power supply variables were evaluated according to a two level factorial experiment. The factors in the experiment were peak current, pulse width, and frequency. Nine equipment settings were used and six replications made at each setting according to a randomized test plan. In the next experiment the peak current was held constant and pulse width and frequency were varied.

Each test combination was evaluated separately. The average flow (sccm) and standard deviation (sigma) were calculated by the Ford time sharing computer "SIGMA" program and the 6 sigma repeatability was determined as a percentage of the average. The best flow uniformity/machining time combination produced a .030" (0.726mm) diameter orifice .060" (1.524mm) long, in 3 seconds with a 6 sigma flow uniformity of 4.5% of mean flow. This was analogous to a hole size consistency of ± .0003" (0.0076mm).

Based on these encouraging results, another series of experiments was conducted to EDM orifices in the nominal range of .01" to .06" (0.25 to 1.52 mm) in zinc die cast material. The investigation was also extended to machine metering slots in aluminum die cast carburetor bodies and a small bleed hole in the carburetor idle tube. Electrode materials, electrode wear and flushing techniques were studied for each application.

The results of this development program yielded the following conclusions:

1. Electrical discharge machining times can compete with current production drilling time.

2. The EDM process has demonstrated machining repeatability as measured by air flow of 4.5% of the mean versus 17.5% obtained with conventional production equipment.

3. Complete elimination of burrs and associated labor costs.

4. Continuous operation can be sustained for at least an eight hour shift with a single electrode.

5. Interfaced with a computer controlled closed loop control system the EDM equipment has the ability to control and vary hole size within narrow limits without an electrode size change.

FORD PRODUCTION EDM APPLICATIONS

METERING SLOTS IN 2V CARBURETOR.

The two idle transfer slots in the main body of the two barrel carburetor had been produced with a conventional punching process. The slots enter into a blind cavity, and the breakout (metal tear) could not be controlled. This contributed to flow variations when compared to each other and caused performance variation from carburetor to carburetor. Slots, .025" to .040" (0.635 to 1.016mm) wide, .140" to .400" (3.556 to 10.16 mm) long through a .070" (1.778mm) wall machined with the EDM process (Fig.1) exhibited improved flow characteristics and less variation from slot to slot.

Figure 1

Experiments also proved that by optimized EDM power supply parameters, proper electrode material and flushing techniques, automated electrode change and semi-automated parts handling, high volume production rates can easily be achieved.

At the conclusion of research and development work and the necessary production engineering, two machines were ordered and production initiated in January 1972.

Each machine (Fig. 2) consists of two horizontal shuttle slides, independently moving forward and back from the operator's position.

Figure 2

Each slide contains a workholding nest and mechanism to automatically locate the part. With this arrangement two carburetor bodies can be machined either simultaneously or independently. When a part is placed into the fixture nest and two palmbuttons are depressed, the machining cycle starts by rough locating the part over two pins in manufacturing holes. The slide shuttles back to a positive stop and a vertical ram lifts the part, positioning the throttle bores over the EDM electrode holders. Locating pins are inserted into the shaft holes and the manifold face is held square against a spring loaded pressure plate. The throttle bores are then plugged and filled with dielectric and the servo controlled forward motion EDM cycle starts.

At this time the horizontal shuttle slide returns to the operator for loading a new part. When the EDM cycle is completed, the vertical ram retracts and places the part into the shuttle slide nest. The slide shuttles back, bringing a new part in position and unloads the completed part on a conveyor.

After each EDM cycle, the electrodes are automatically adjusted to compensate for wear by movement against a fixed stop. When the electrodes wear to a predetermined length, a new electrode is automatically inserted into the holder from a self contained storage magazine containing 50 electrodes.

Two dual 25 A output EDM power supplies are used with quick changeover features to facilitate mainenance, and the sequential functions for both machines are directed by solid state controllers.

SIDE BLEED HOLE IN CARBURETOR IDLE TUBE

The uniformity of the side bleed hole in the carburetor idle tube has a significant effect on the idle flow characteristics of a carburetor. The tube (Fig. 3) is hard drawn brass material, with hole size of .026" ± .001" (.660 ± 0.025mm).

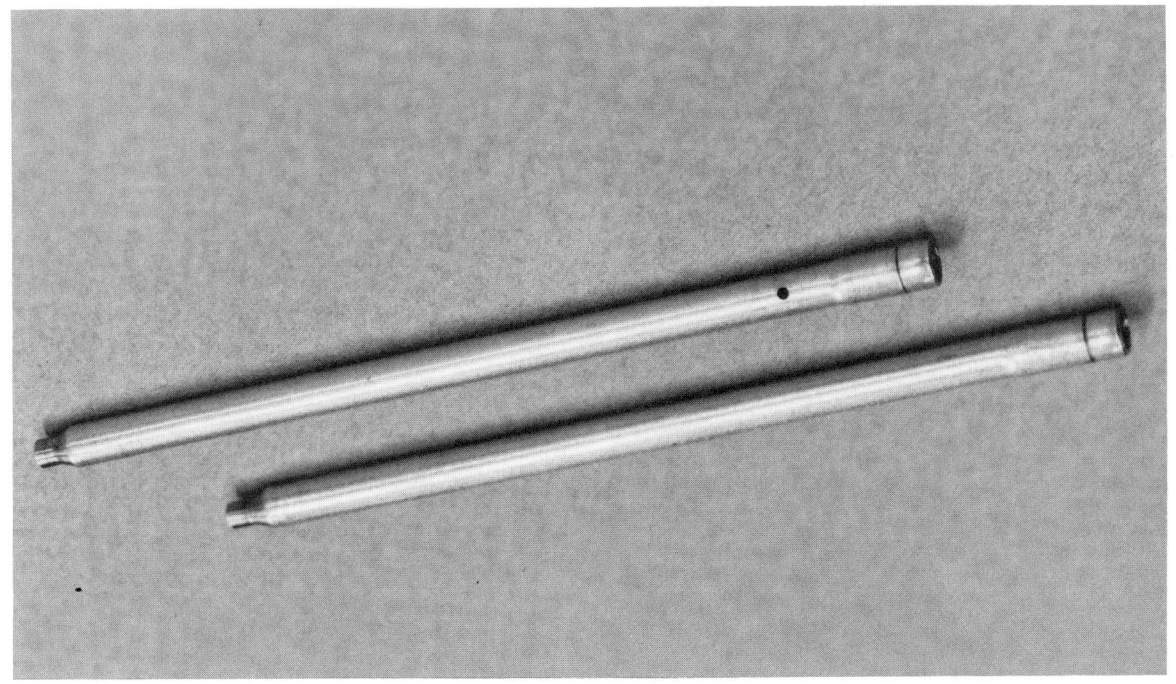

Figure 3 (3.5X)

Conventional drilling invariably leaves a burr on the inside of the tube, contributing to flow variations. Deburring is difficult and costly. To eliminate these problems, it was decided to EDM the hole. Due to high volume production requirements (several million pieces per year), specially designed EDM equipment was needed to reliably maintain production. The process as developed is fully automated, and production was started in September 1971.

The basic machine (Fig. 4) consists of a precision work base with a table top keyed to locate work fixtures, a vertical column assembly, a removable cartridge type EDM head mounted to the vertical slide, a hopper feeding mechanism and approach and ejection tracts. A hydraulically actuated shuttle work-holding fixture with precision ground inserts to nest the tubes in 'V' grooves completes the assembly.

Figure 4

The automatic vibratory feeder hopper orients and positions the tubes, allowing approximately 200 tubes to be in the approach track at all times. The tubes travel through the track on two parallel, hardened runners. The shuttle fixture nest picks up 10 tubes, positions them under the EDM station and holds them in position by spring loaded urethane pressure pads. Upon completing the EDM cycle, the shuttle fixture will release the finished tubes and pick up the next 10 tubes to reposition under the EDM station, pushing the machined tubes through an escape track.

To facilitate electrode change, the EDM head is a removable quick disconnect cartridge assembly, fitted with 10 independent continuous electrode mechanisms and automatic electrode wear compensator.

The EDM power supply is a multi-electrode 10 channel output unit with 0.5 A output per channel.

Silicon dielectric is used and is recirculated through a conventional filtration unit. The electro-hydraulic servo system is solid state controlled.

HIGH SPEED BLEED HOLES IN CARBURETOR BOOSTER VENTURI SUPPORT

As mentioned earlier, experiments proved that metering holes produced with EDM exhibited less flow variation than the conventional process used. It was decided, therefore, that the high speed bleed holes in the carburetor booster venturi support (Fig. 5) .052" ± .001" (1.320 ± 0.025mm) diameter would be produced by EDM to assure flow uniformity needed for emission control. The fully automated process was started in January 1972.

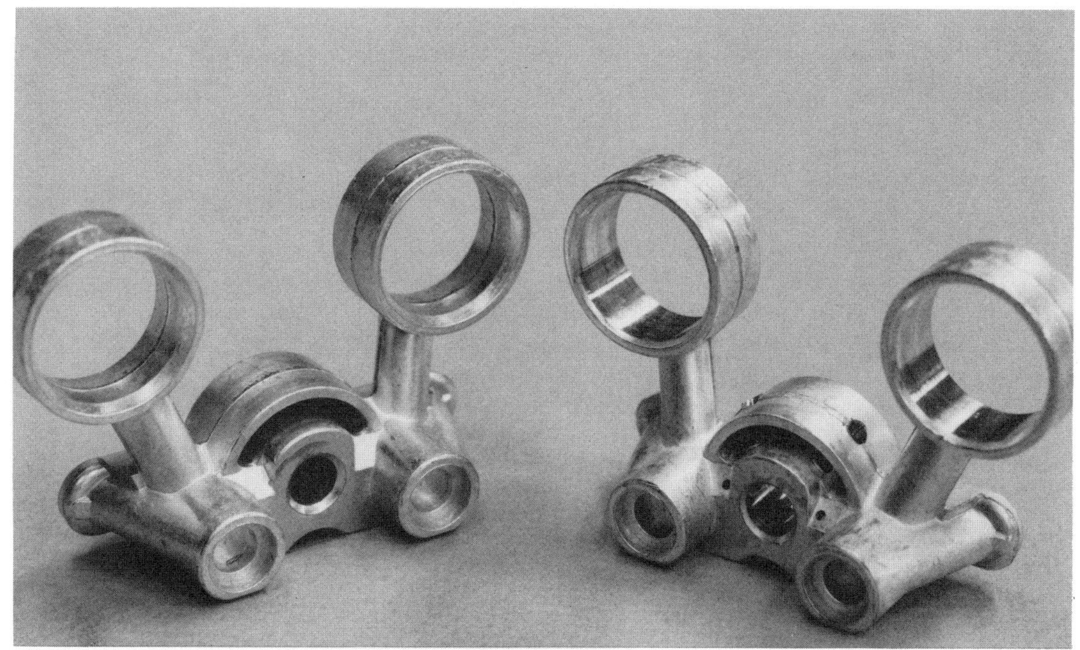

Figure 5

The machining is performed on a 12 station center-column rotary machine. The parts are auto loaded into a dual fixture and after conventional drilling, reaming and deburring operations, the high speed bleed EDM holes are produced on stations 9 and 10. The parts are auto unloaded for further processing.

The two EDM heads (Fig. 6) are of identical design with carriage assemblies capable of machining two holes simultaneously, using two independent refeed mechanisms for electrode wear compensation.

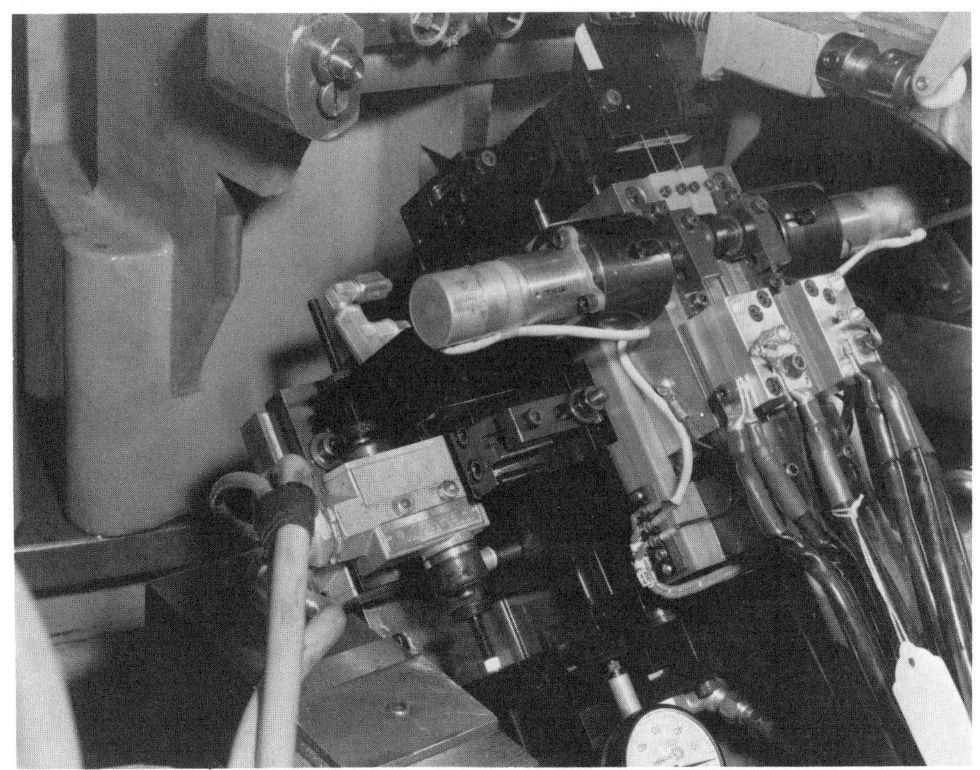

Figure 6

Each EDM head is fed by a dual 25 A power supply, modified for closed loop servo control of pulse on time.

To make the EDM operation compatible with conventional metal cutting operations on the same machine, a first in the industry, a solution had to be found for providing coolant for conventional machining and dielectric for the EDM stations. A simple but ingenious solution was adopted. Instead of water soluble coolant, EDM dielectric fluid is used on all stations as coolant, with excellent results. No deterioration of machining quality was observed. A conventional chip removal and filtering system is used, with additional filtering for the EDM stations. The dielectric fluid is continuously recirculated.

To assure the uniform flow characteristics of the holes, a hot film anemometer air flow test stand was incorporated in the system and interfaced with a computer. In the test cycle the part is loaded into the test stand fixture and the two holes are tested at the same time individually. The flow (sccm) is calculated by converting the analog voltage readings from the flowmeters as follows:

A standard calibration curve of the flowmeters is loaded in the computer memory, consisting of flow values of the range to be measured at a given pressure differential at standard atmospheric conditions, 70° F and 29.92" Hg (21.1° C and 760mm Hg).

Three reference orifices are provided to establish a flow curve under present conditions, and frequently updated to compensate for changes in ambient conditions.

The computer will monitor the flow values of the holes and compares the reading with the specification mean flow value. After a minimum of 5 samples have been tested, the operator initiates an "Action" button on the test stand. If the average air flow of the holes is within .5% of the specification mean, no corrective action is performed and production continues. If the average deviates from the specification mean by .5% or greater, the computer calculates the magnitude of the adjustment required to correct the flow and outputs the proper number of pulses and directional signal to a servo motor system to change the EDM power supply pulse on time. This operation is interlocked with the main machine cycle, so no EDM operation will be started before the adjustment is completed.

Additional interlocks are also provided to detect severe malfunctions. If the average flow deviates significantly from the specification mean, an alarm is printed out and production halted until corrective action is taken and additional parts tested by use of an over-ride system. If the two orifices within one part flow significantly different, a similar alarm is triggered to initiate corrective action.

CONCLUSION

Monitoring the flow characteristics of a hole, instead of its mechanical dimensions, is a radical departure from standard EDM practices. Flow test takes into consideration all parameters affecting the flow; hole diameter and length, hole inlet and outlet edge conditions and inside surface finish. All conditions being optimized, EDM is a very stable process, therefore, only minor adjustments are needed to compensate for small variations in electrode diameter, part wall thickness or material density. It has been found that by automatically adjusting the pulse on time within narrow limits, a positive control of flow magnitude can be achieved. To the author's knowledge, this is the first time in industry that an EDM process has been brought under the control of a computer.

ACKNOWLEDGMENT

The development of this process was a joint effort of the Ford Motor Company Manufacturing Development Office, General Products Division Staff Advanced Manufacturing Engineering and Rawsonville Plant Manufacturing Engineering. The contribution of each and every individual participating in the successful completion of this project is gratefully acknowledged.

CHAPTER 5

WIRE EDM

Wire EDM

By Hans R. Gerber
Territory Manager, Charmilles Corporation of America

I. Introduction to EDM and Wire EDM

EDM, electrical discharge machining, is a process using non-stationary, time-spaced electrical discharges between an electrode and a workpiece to remove material.

EDM has two basic capabilities, which are:

a) Machining of any materials that are electrical conductors --hard or soft-

b) To reproduce any shape, 2 or 3 dimensional, without relative movement between electrode and workpiece other than vertical penetration.

Spark erosion can be divided into three categories:

1. Sinking by EDM

   a) 2 dimensional die sinking

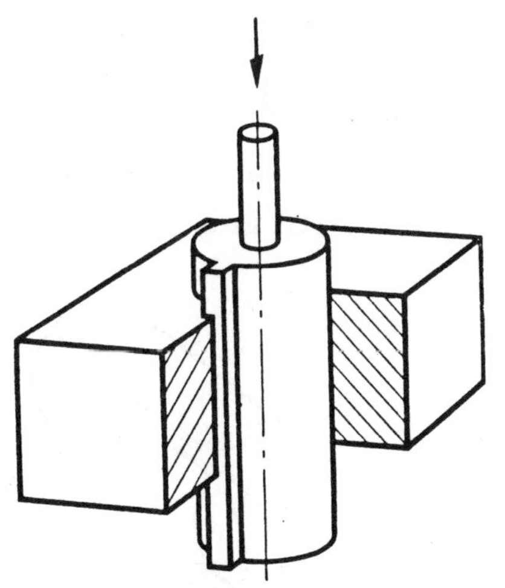

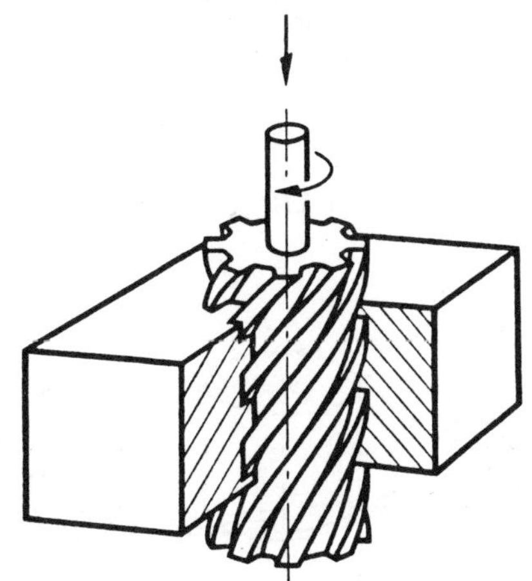

b) 3 dimensional cavity sinking

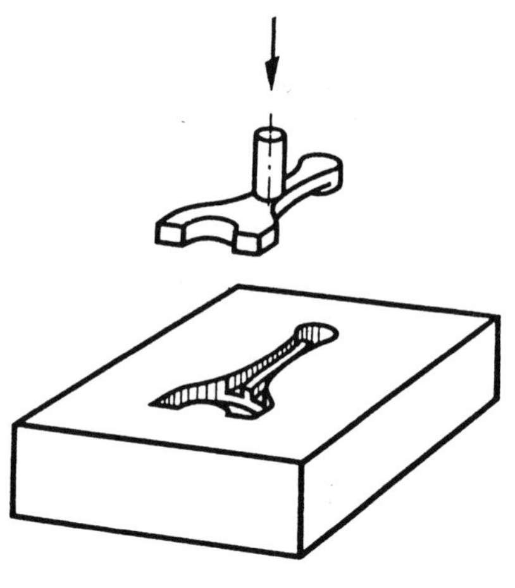

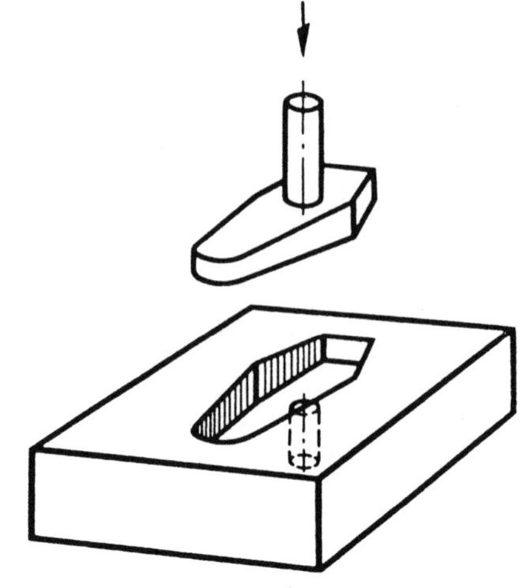

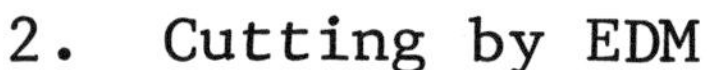

2. Cutting by EDM

a) cutting with a blade

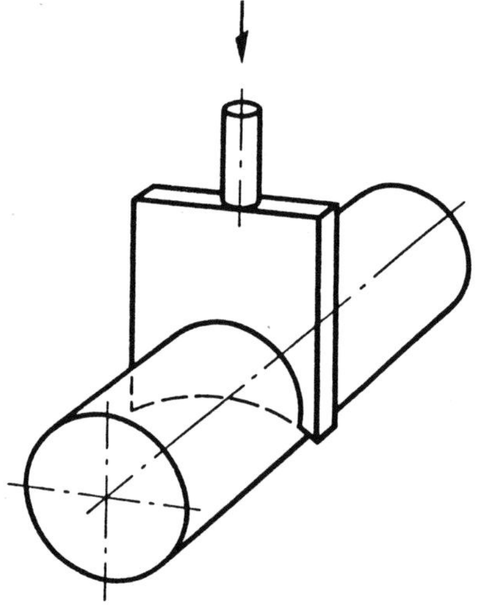

b) cutting with a wire

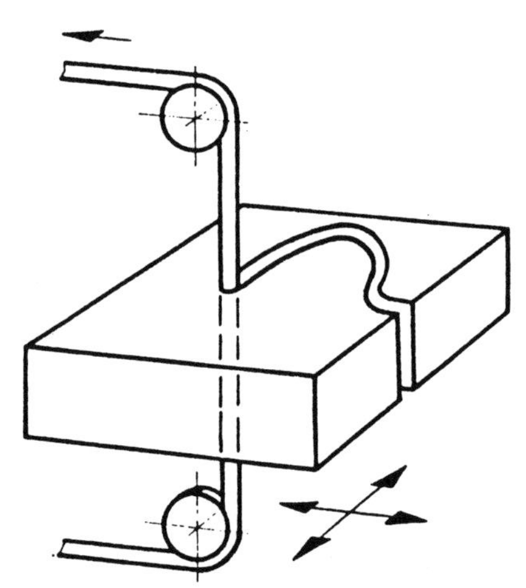

c) cutting with a ribbon

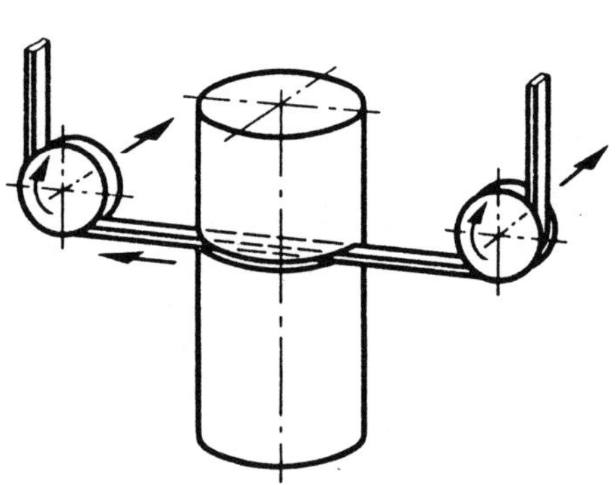

d) cutting with a disc

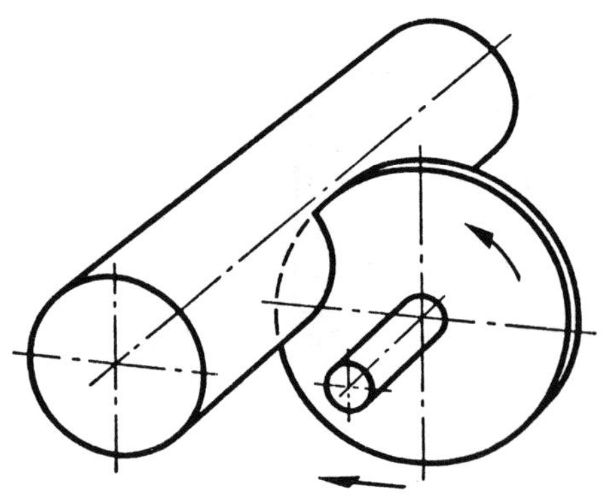

3. Grinding by EDM

a) external grinding

b) internal grinding

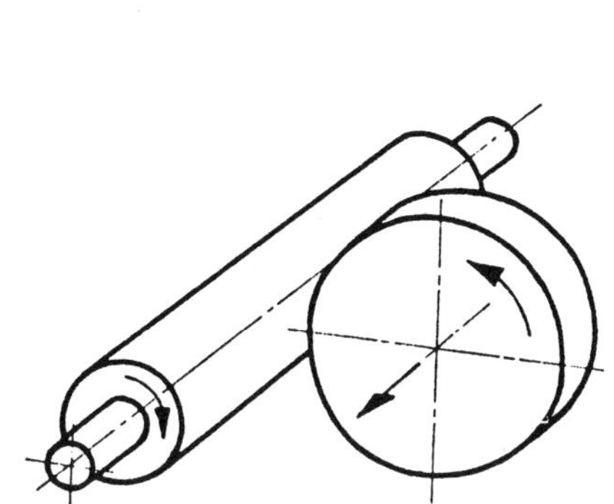

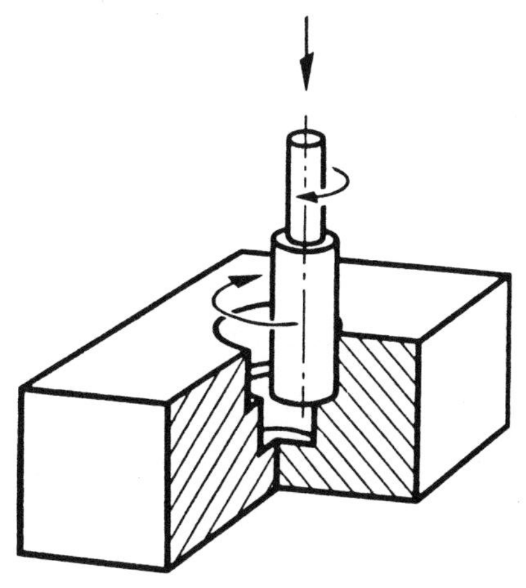

c) surface grinding (flat or contour)

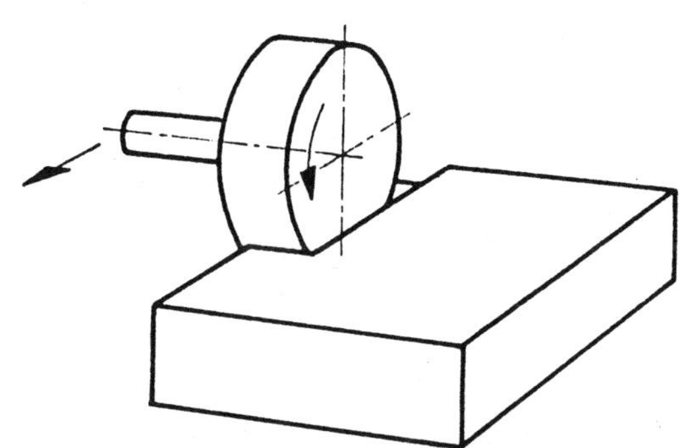

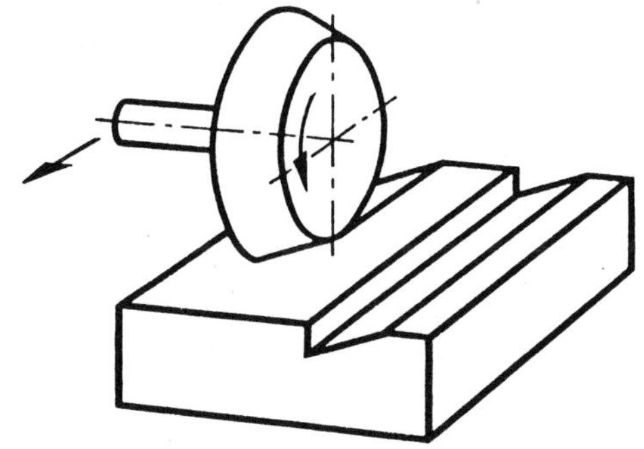

The ram type machine, fitted with the proper electrodes, can produce most of the operations in 1 to 3. One exception is the EDM surface grinder used for 3c.

In standard EDM work, manufacturing of the electrodes usually represents an estimated 40% to 60% of the total cost of the EDM work. Due to these electrode costs, a new type of EDM machine has been developed: the travelling wire machine, (see 2b). As its name implies, the electrode is a thin wire (the actual tool) which renews itself by unreeling from a spool. After having done the work in the cutting zone, this wire is rewound and discharged. Crudely said: "We have an EDM bandsaw". The movement of the workpiece in X and/or Y axis is achieved by using electronic drive systems. The controls are NC, CNC or DCNC.

The apparent limitation for this type machine is that only through-hole work can be done.

The advantages are:

1. No electrode has to be made;

2. Tapes for die shapes can be reused over and over again;

3. The workpiece materials are the same as for standard EDM; that means all conductive materials--hard or soft, metallic or nonmetallic.

II. Wire EDM

A. Machines

The machine tool for wire EDM has the same function as the standard EDM, mainly:

1. Holding the workpiece;

2. Holding the electrode;

3. Moving the electrode into the workpiece.

Design and construction of a wire EDM system can vary considerably. Priorities like price, size, or guaranteed accuracies will have an effect. Therefore, various machine tool constructions are available as follows:

1. The wire-holding device is fixed on the machine head and X-Y table is used to move the workpiece.

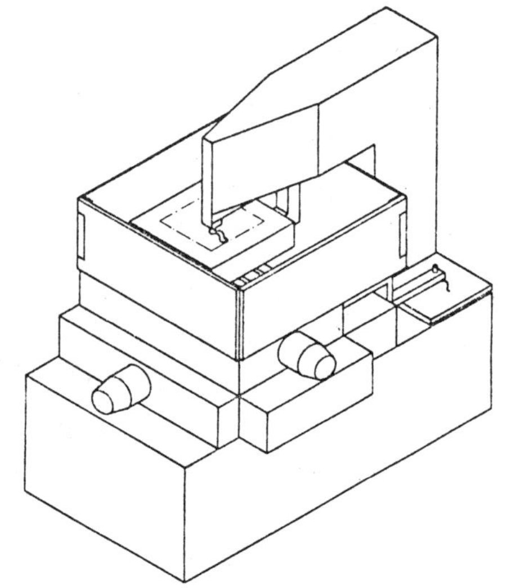

2. Same as #1 but different design.

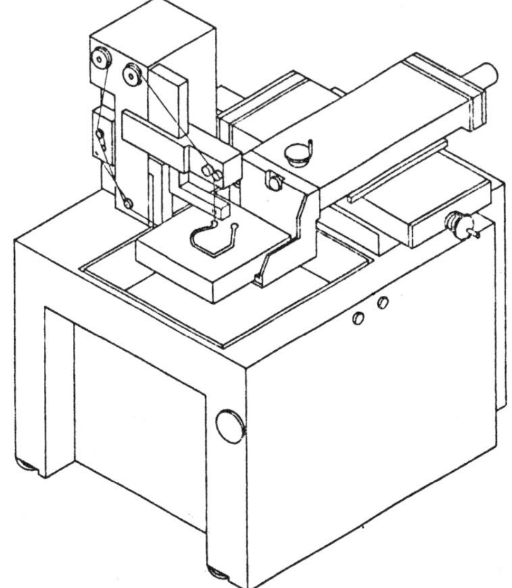

3. The wire carriage system moves in X axis, the table supporting the workpiece in the Y axis. This is a bridge type design

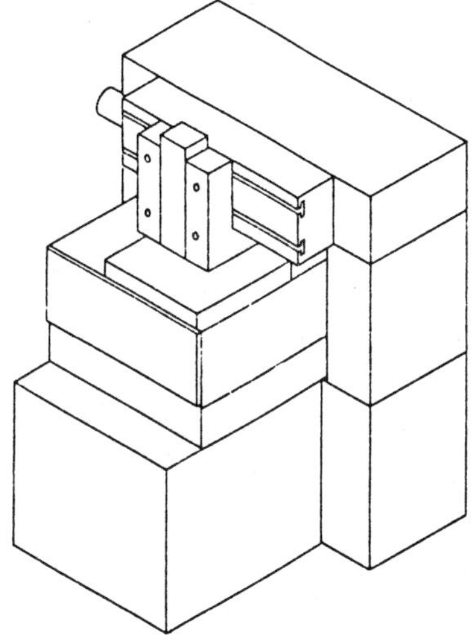

The maximum usable weight of a workpiece which can be held on the machine tool will be influenced by the way the machine is designed (1, 2 or 3), and what accuracy is guaranteed.

Table Movement

The X or Y axis movement is achieved by using either DC motors or stepping motors.

Three different drive systems are used today:

1. A stepping motor which drives the lead screw and displaces the table or the head. This is called an open loop system:

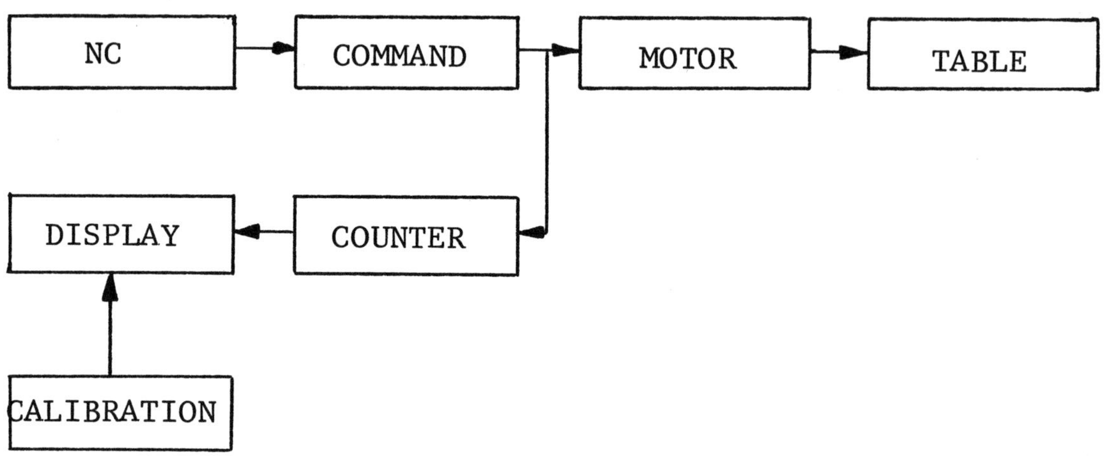

The system using stepping motors requires a precision lead screw and does not tolerate motion resistance (such as experienced in milling machines, etc.).

2. Drive with DC motors with semi-closed loop system which measures the displacement of the lead screw in the angular position. This method also needs a very accurate lead screw, frictionless moving members, with no forces working against the movements:

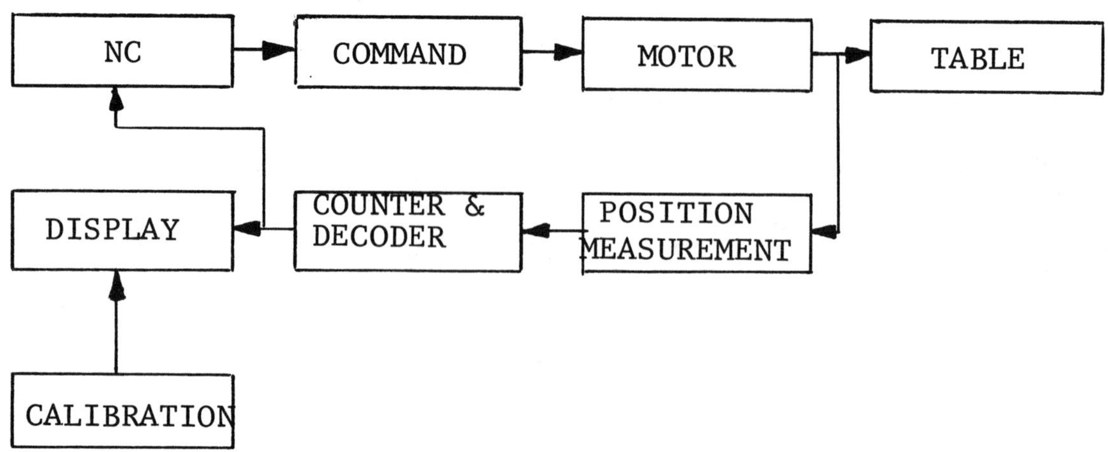

3. The last system is the closed loop system which measures every step taken, and feeds the information back to the NC control. With this method the lead screw does not have to be accurate and heavy forces can be applied to the table:

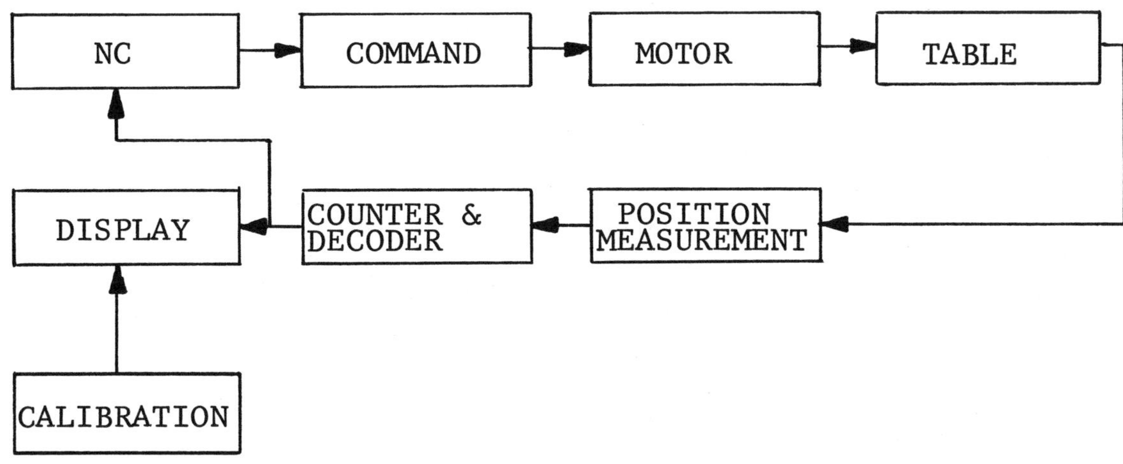

Most of today's wire machines use the first two systems. The stepping motor gives us the possibilities of achieving linear displacements of .000040" per step on some and .0001" per step on other machines.

B. Wire Transport and Guiding System

The wire feed mechanism is another very important part of the machine. Since accuracy requirements on the finished tool are usually high, the same is true of the wire feed mechanism.

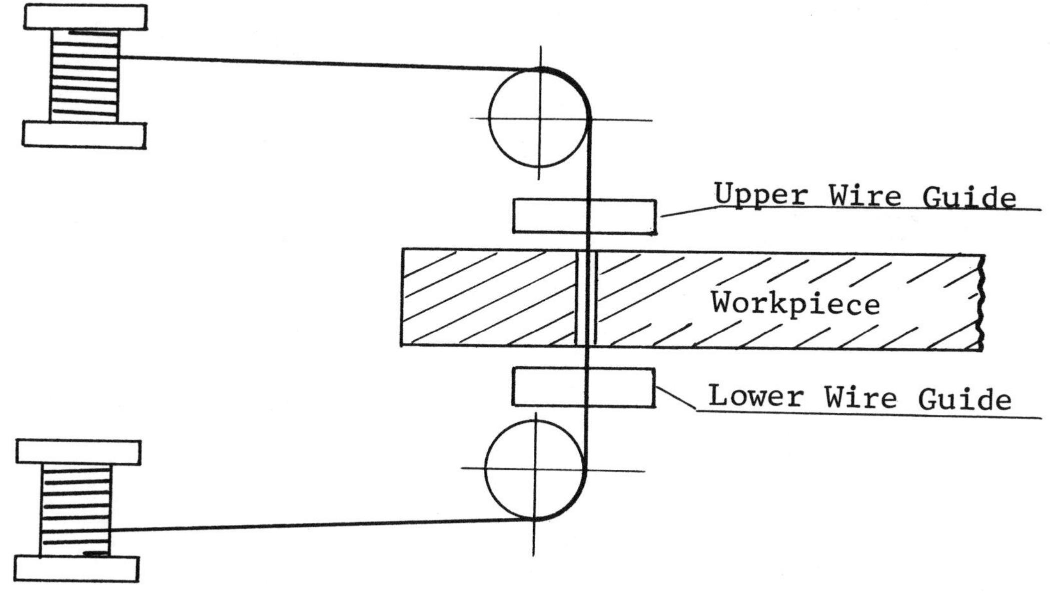

The ultimate feed mechanism has a wire with a constant diameter, moving through the workpiece at constant unreeling speed in an exact vertical plane, with absolutely no side motion of the wire or wire vibrations.

Complex systems are employed by the various manufacturers. There are different wire materials available. The most commonly used:

002"-012" diameter copper or brass wire

003" diameter or less - molybdenum wire

The material is purchased wound on a spool of approximately 4-8 lbs. If untreated, it will tend to retain its original bent form while moving through the cutting area. It will not be perfectly straight thus affecting finished part accuracy. Here are some techniques used to "straighten out" this wire.

1. The wire is pulled through a series of roller straighteners before going to the working area.

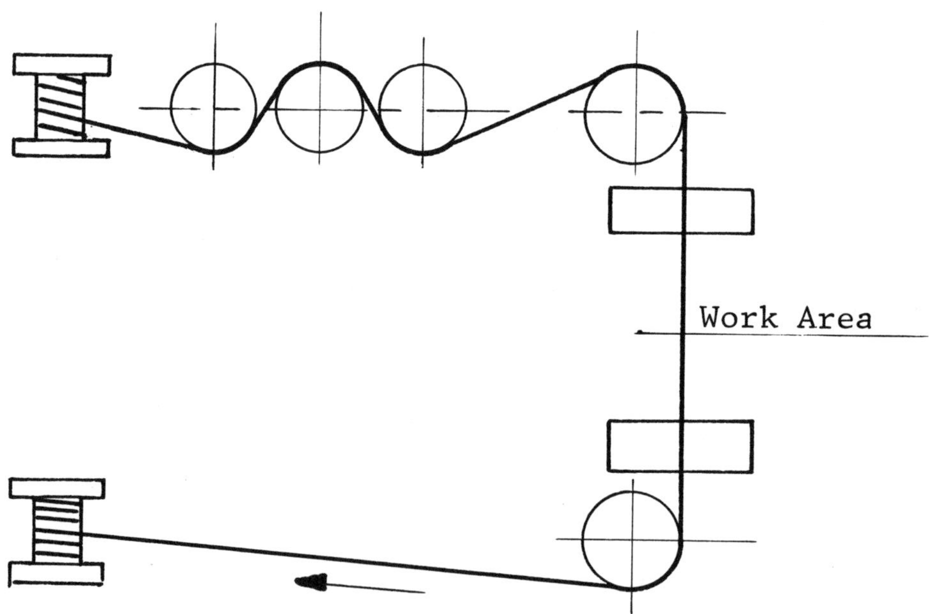

2. The wire is moved through two drive pulleys on the same shaft. The second one is slightly larger than the first. Through this speed differential, the wire will be stretched past its elasticity point and loses the memory from the

spool. No sharp bends should occur in the wire path until after the cutting zone.

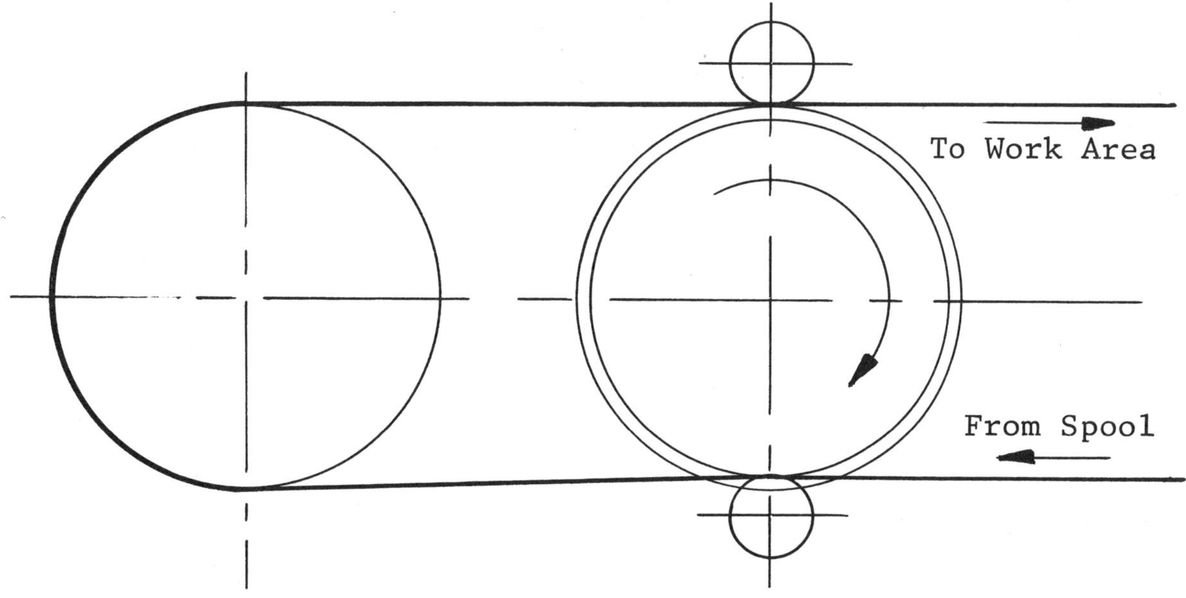

3. One system favors brass wire as the electrode. This wire is pulled through a drawing die and is precisely calibrated right on the machine. The straightening method used is a thermic one. The wire is guided through an annealing zone in the vertical axis directly above the working zone.

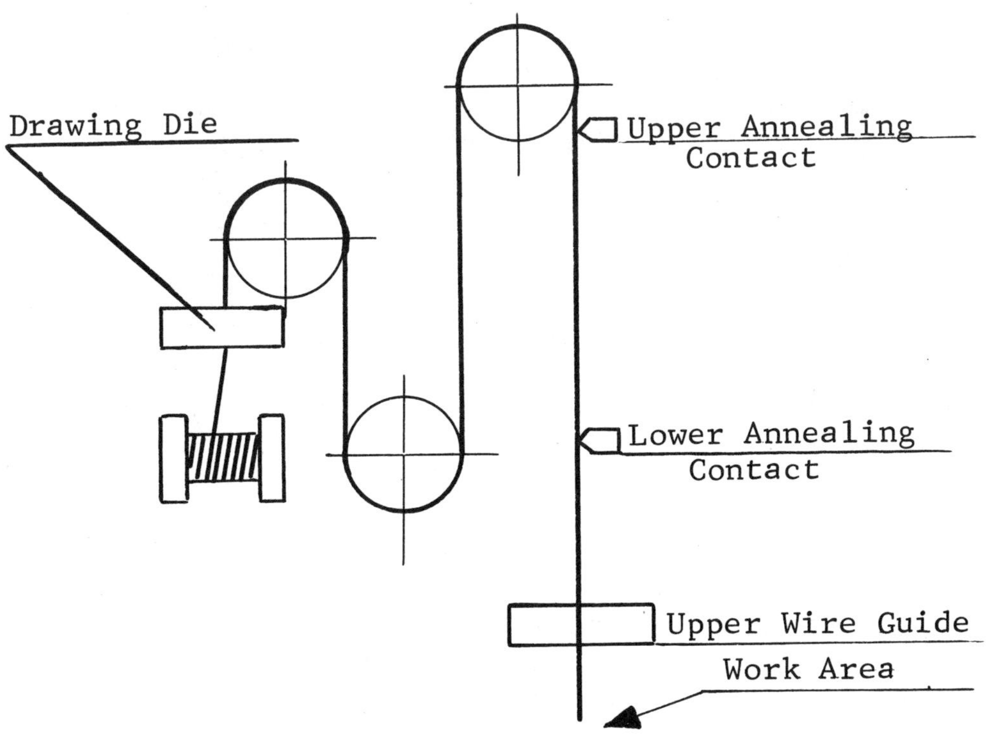

The guiding system of the wire in the work zone has to be very accurate and free of wear on the contact points. Sapphire is used in V-form for this purpose.

Direct taper cutting can be achieved by offsetting the upper or lower guide and orientating it tangentially to the direction of the cut. This movement is computer controlled by means of a stepping motor.

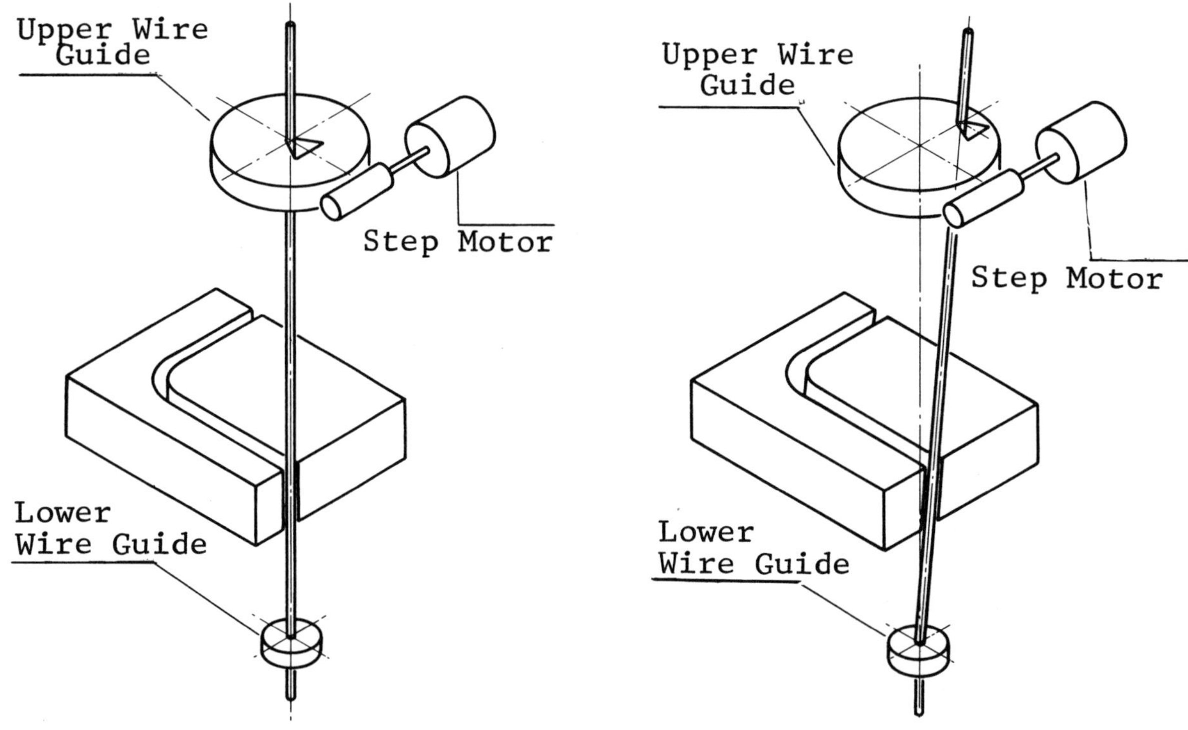

There are mechanical methods of achieving tapers in a die by:

1. Mounting an aluminum or steel plate on insulator above the workpiece. Both plates are cut straight at the same time. Then remove the electrical connection between plates and

dial in a larger offset. On the second pass the wire will be dragged along the upper shape (used as template) and will cut a taper.

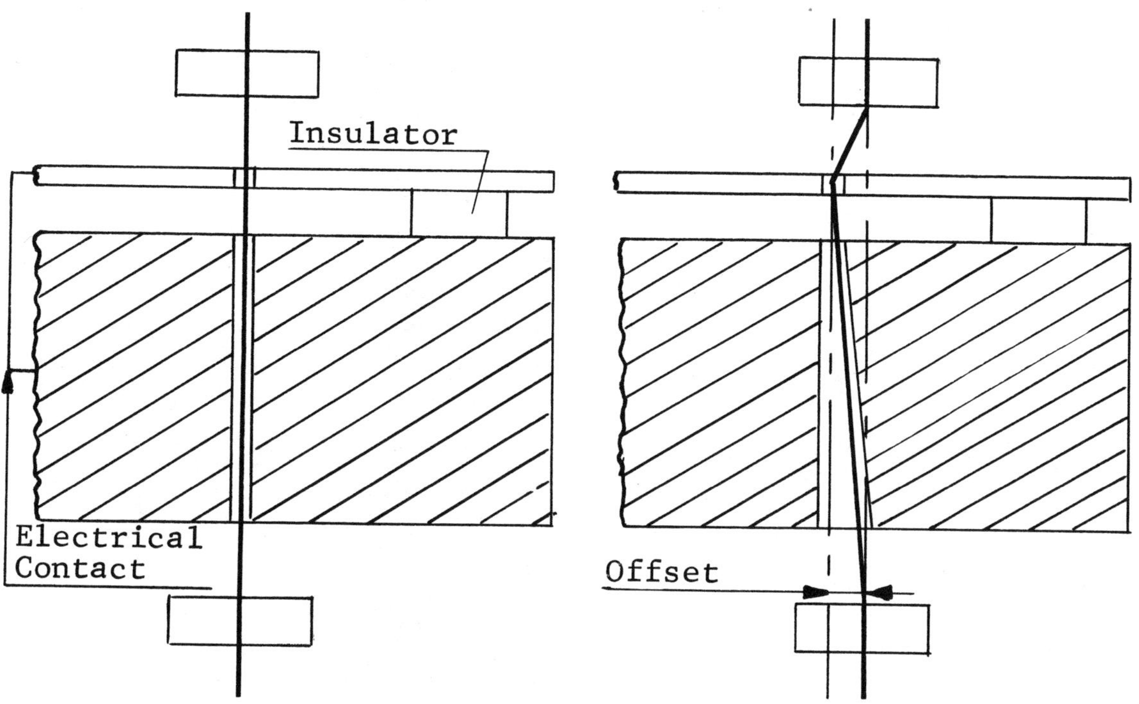

2. After making a straight cut, a larger offset and a plastic bushing are used to mechanically guide the wire around the upper edge of the die.

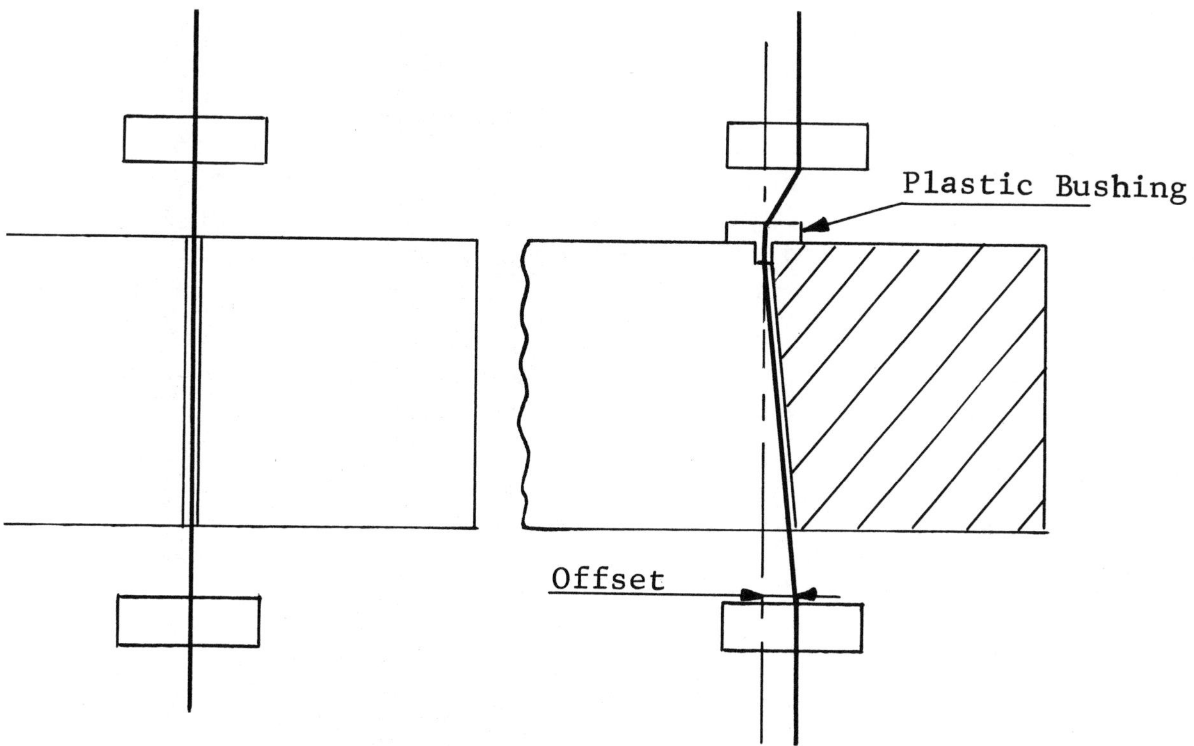

C. Dielectric System

Unlike standard EDM, water is used in wire machines. Mineral salt can change the dielectric strength and therefore change cutting speeds and overcuts. The water is deionized and demineralized through chemical filtering for performance. In addition, regular filters are used to trap sludge from workpiece and electrode erosions.

Various flushing systems are used:

1. The workpiece is completely submerged in dielectric which helps the thermal stabilization of both workpiece and machine tool. A light, adjustable flow of water over top and bottom of workpiece will wash the sludge away.

2. Flushing jets are directed into the cutting zone only without submerging the workpiece. Adjustment of the flushing is very critical with this method, and on long running jobs rusting of the workpiece may occur.

D. Generators

Generators or power supplies produce the energy waveforms which will provide the sparks or discharges between wire and workpiece. This will, in turn, remove the metal and a change in gap voltage will allow the NC or computer unit to move the table forward. Generators employ various principles to produce the discharges in a wire machine. Due to the small contact area - wire-workpiece - only low energy discharges can be used. Usually a high voltage with a short discharge duration is employed. It can be produced through transistorized or capacitor type circuitry. If the average voltage can be lowered to almost zero, the electrolysis effect in the gap can be eliminated and no plating effect between wire and workpiece takes place.

E. NC, CNC, DCNC

Each manufacturer uses a different system of driving the X, Y, and Z axes.

NC Machines

On an NC machine the tape to drive the unit must be made with the help of a computer link-up. Such a unit (in-house or time sharing) will produce a tape for a particular job.

DCNC Machines

These units use their own minicomputer to drive the machine axis. The computer can be programmed, similar to the large outside units, and will run the machine directly without using an intermediate step of a tape.

III. Programming

There are a few possibilities for programming the wire machine. On NC and CNC very simple shapes can be programmed directly by figuring all coordinates with the help of a calculator. For more complex jobs such as a gear die, or a lamination die, the use of a computer is needed. This can be in the form of a link-up with a large computer such as MDSI uses or with an in-house minicomputer such as a Hewlett-Packard. This system is used to produce the long tape needed to run the NC drive.

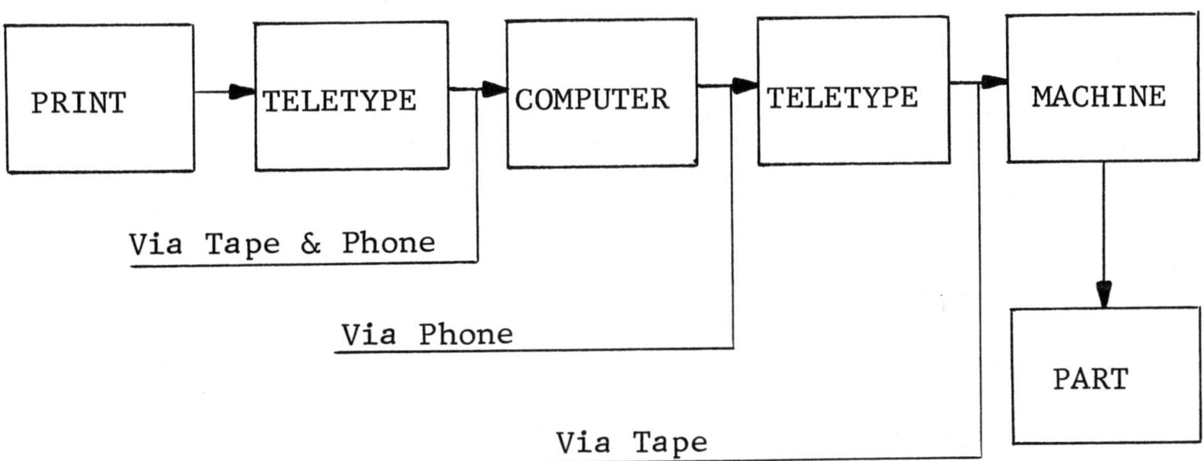

On the DCNC machine the minicomputer is built-in and drives it directly from a memorized program. A shop drawing is needed to program the computer geometrically defining the circles or lines in sequence. All other mathematical functions, to figure out all points in between, are carried out and stored in memory by the computer until this information is retrieved to be passed on to the stepping motors.

This system enables the operator to ask the computer for a printout of all pertinent data for all lines, arcs, intersection points not available at the time of programming. Axis shifts and scale changes (including decimal), up-and-down from the original size, can be done without having to change the original tape, by simply addressing the computer and giving the additional data before refeeding the tape.

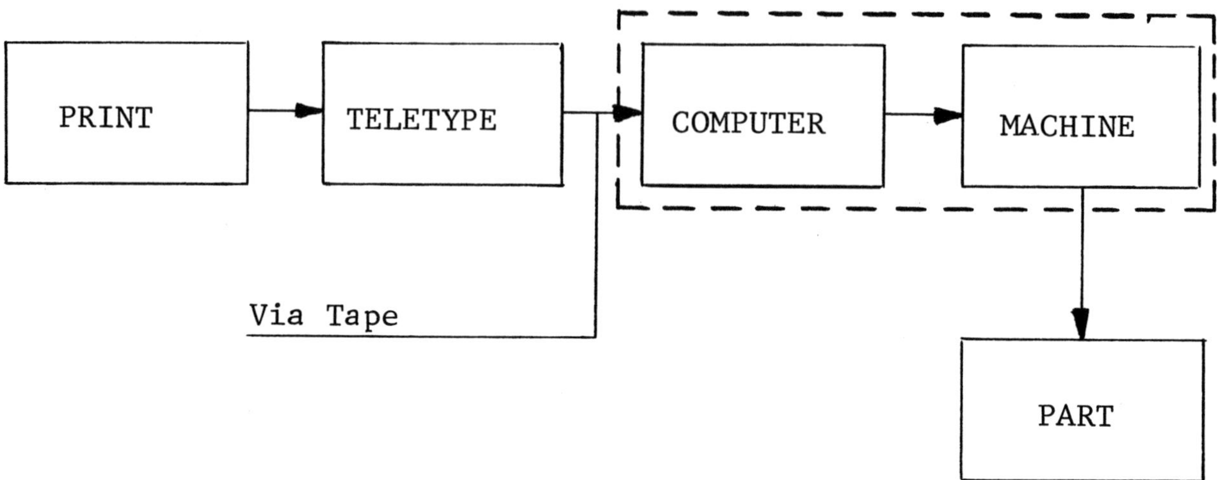

In addition, the CNC and DCNC machines use their built-in computer to orientate the upper wire guide. This enables the user to cut the openings with built-in taper in a single cut. (see wire feed mechanism)

IV. Practical Use of a Wire Machine

A. Advantages

In this form of "EDMing" no electrodes have to be made which, --as many users of standard EDM know--cost money, need skills and equipment not to mention the cost of electrode materials. The average electrode cost on a wire machine for one hour of operation runs approximately $0.20. On thin steel pieces (1/2") the electrode cost is low in contrast to 3" to 4" carbide which commands higher wire unreeling speeds resulting in higher electrode cost.

Machining speed on this type of equipment is expressed in mm^2/min. or square inches/hour. The cutting rates of these machines range from 2-35mm^2/min., depending on workpiece material, thickness and finish requirements. The width of the slot is completely disregarded for job estimates since the wire diameter remains contant. In order to predetermine the cutting time the periphery is measured and divided by the cutting speed. Here lies one of the major advantages of the wire process: the linear speed is the same whether the machines erode a 5" straight line or the 5" periphery of a gear die. This shows that wire EDM will have tremendous savings on complicated parts such as lamination dies, gear dies, etc.

Prototype parts can be cut very economically on these machines. The flat stock is welded together on the outside corners and the parts cut out all together from the center section. These pieces can now be evaluated and a die be made with the same tape.

B. Disadvantages

Unfortunately, every advantage has a drawback or disadvantage. In wire EDM it is price. Standard EDM machines cost from $15,000 to $40,000. On the other hand, the wire systems today run from approximately $90,000 to $190,000 depending on size, quality, capability, accuracy and sophistication of the system.

In case of larger die openings made in solid steel the material should always be stress relieved. Otherwise, a distortion will take place when the large center portion is cut out of the workpiece. A second cut, called whipping cut, could be used to compensate for any such distortion. Another method would be to have the plate preroughed before heat treating.

For every inside shape a prehole or starterhole must be available so the wire can be threaded through. This can present a manufacturing problem if the prehole has to be .020" diameter or smaller (in small sections).

C. Wire or Standard EDM

Standard EDM for 3-dimensional cavities such as forging dies, plastic molds, coining dies, jewelry dies, etc., is unchallenged. In 2-dimensional openings such as stamping dies, fine-blanking dies, powder metal dies, and extrusion dies, wire EDM is a strong challenge to standard EDM. If the opening requires expensive electrodes, wire definitely has the advantage. Simple round or square openings, with low electrode cost, still belong on standard EDM.

D. Wire EDM Only

In many cases wire has opened new fields for EDM. Whenever an extrusion die requires electrodes with thin cross sections and/or considerable length, it is almost impossible to make electrodes. The wire EDM machine now can make the die or the male electrode depending on which procedure is more economical.

V. Summary

As described in the previous pages, one can see that the WIRE SYSTEM is a logical addition to EDM. It will not replace all the standard machines. The estimate is that about 5%-20% of the work done on standard machines can, and eventually will, be done with this "NEW" process. But even more, wire opens new fields for EDM in metalworking, and will permit users to select tougher materials, adopt new designs, etc.

N/C EDM—"Cutting" Costs With N/C And A Wire Electrode

By C. Michael Mercincavage
Senior Associate Engineer, IBM Corporation

The introduction of Numerically Controlled Electro Discharge Machining (N/C EDM) was about as well received as a draft notice during wartime. The idea of buying a piece of shop equipment incorporating the N/C EDM concept would have raised critical questions about ones' professional competence and state of sanity.

Fortunately, the early attitudes have changed. N/C EDM is finding its way into progressive thinking shops and paying off in significant savings when used for fabrication of tooling components, pressworking tools, miniature parts of complex profile, erosion of parts from exotic materials and, in many ways, resulting in a competitive advantage over other methods. The versatility and flexibility of the process will undoubtedly make N/C EDM a necessary addition to any shop expecting to remain competitive in the tool and die, job shop and production operation arena.

Figure 1 is a graphic representation of the basic N/C EDM process. The strength of N/C EDM lies in the basic makeup of the process. To the advantages of EDM, add the advantages of Numerical Control.

First, consider the general advantages of EDM. Some of the advantages include: The ability to erode hardened-tool steels, tungsten carbide, and some of the exotic materials with good finish and tolerance control. Also, consider the minimal fixturing requirements resulting from the fact that cutting forces are negligible. Further consider the in-cycle free time, available to an EDM operator, while a job is running. There are obvious advantages in an efficient arrangement that permits running of another machine tool in-cycle to an EDM operation. This situation is particularly true with N/C EDM.

Secondly, consider the advantages of N/C. At the top of the list is the inherent flexibility resulting from programmed sequences of instructions. New jobs can be set up and run with only minor changes (usually only tooling changes). Job runs may be stopped and started at specific program locations as desired. After a first piece is approved, repeatability and accuracy are consistent unless the operator intervenes or program changes are made. Because the machine tool functions are controlled by the Numerical Unit, a job may be run with only minimal operator attention.

The combination of features from N/C and EDM result in a process that will machine almost any material which is electrically conductive at room temperature. The process will produce two-dimensional shapes that are limited in profile complexity only by programming capabilities, and certain restrictions imposed by workpiece size. The former limitations are minimal if state-of-the-art programming languages are employed. Once set up, a job may be left unattended during the erosion

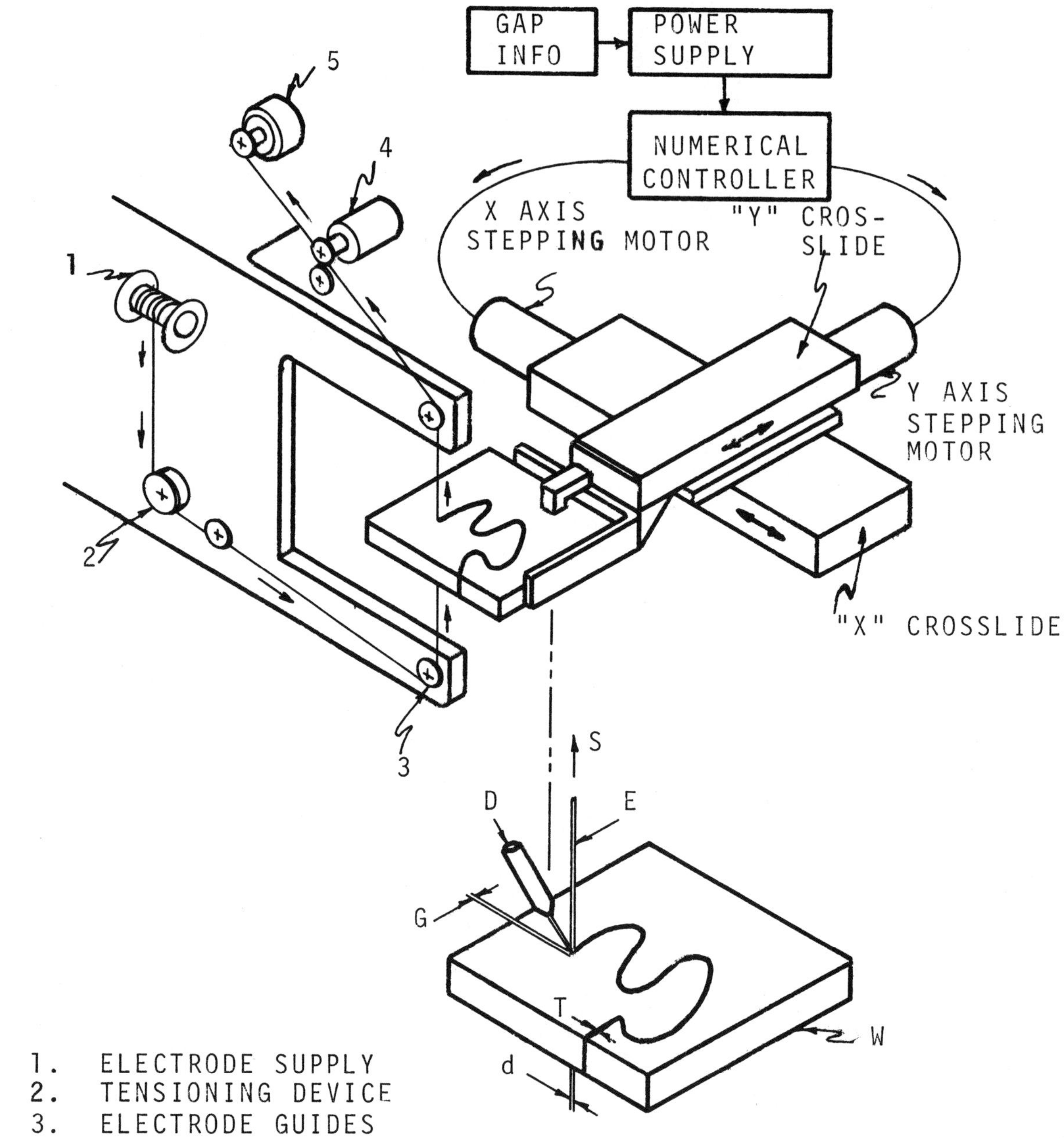

GAP INFO

POWER SUPPLY

NUMERICAL CONTROLLER

X AXIS STEPPING MOTOR

"Y" CROS-SLIDE

Y AXIS STEPPING MOTOR

"X" CROSSLIDE

1 2 3 4 5

S D E G T W d

1. ELECTRODE SUPPLY
2. TENSIONING DEVICE
3. ELECTRODE GUIDES
4. ELECTRODE DRIVE
5. TAKE-UP SPOOL

S=WIRE TRANSPORT SPEED
E=ELECTRODE WIRE
G=SPARK GAP
T=TRACK WIDTH
d=ELECTRODE DIAMETER
D=DIELECTRIC
W=WORK PIECE

FIG. 1
THE BASIC N/C EDM PROCESS

cycle. This combination of advantages offered by N/C EDM are best illustrated by the following example:

> Assume that you are required to produce a die plate, and two punches for a repress die with a complex contour, (with several tangent radii, points of inflection and fillets of 0.01-inch radius in the profile as shown in Figure 2.) The die plate is to be of 2.0-inch thick tungsten carbide and made in one piece (no sectioning). It's Friday morning, and the job must be completed by Tuesday morning. Completion of this order on the due date could result in a substantial follow-up order.

This is one of those situations where you wish it were possible to share the enthusiasm and confidence of your sales staff. But experience dictates otherwise. You know that the only practical way to do the job is with conventional EDM. The first big snag is fabrication of the electrodes; several roughers and finishers would be required for both the die plate and punches. Finding the free time or making some time available is a problem in itself, especially since the situation will call for substantial direct-labor overtime. After the electrodes are fabricated, possibly by late Saturday afternoon, the next problem would be soliciting an operator to monitor the EDM roughing operation and to change the electrodes as the erosion progressed. When this portion is completed, the finishing operation will be required. To minimize taper, frequent electrode replacements will be necessary. Assuming that the die opening was cored by a cylindrical electrode to minimize erosion demands on the shaped electrodes; the cutting times would be in the range of 8-12 hours as a minimum. Next the punches must be fabricated. Despite the fact that we are not going into all the details of manufacture, some hard, cold, expensive facts are already beginning to surface. First, is the fabrication of electrodes and second, is the requirement for direct-labor support on both a regular and overtime basis. Both of these factors, the electrode fabrication and direct labor, could be eliminated or minimize by N/C EDM. Figure 3 shows the estimated costs, in terms of direct labor, for conventional fabrication.

Consider the same job again, only this time assume your shop has an N/C EDM machine available for use on the job. Three basic requirements must be met to begin the job:

> First, machine-tool time must be available.
>
> Second, and N/C tape must be punched. This single tape will contain the information to describe the punch and die profile and the necessary functions for running or stopping.
>
> Third, an operator must be available to setup the work-

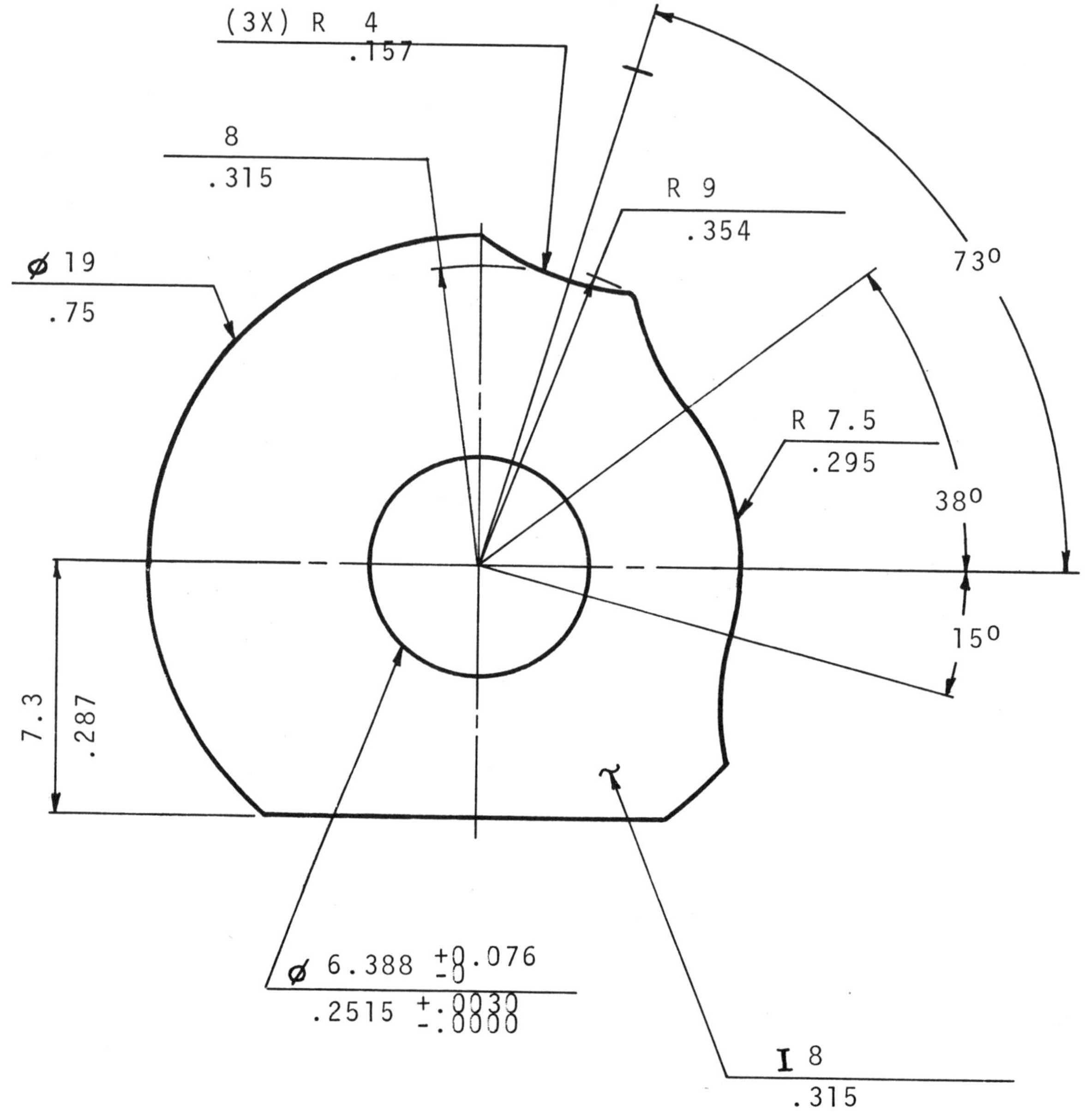

(3X) R 4
.157

8
.315

R 9
.354

73°

Ø 19
.75

R 7.5
.295

38°

15°

7.3
.287

Ø 6.388 $^{+0.076}_{-0}$
.2515 $^{+.0030}_{-.0000}$

I 8
.315

FIG. 2
REPRESS DIE CONTOUR

JOB DESCRIPTION/ OPERATION	HOURS CONVENTIONAL		HOURS N/C EDM		N/C EDM SAVINGS/(COST)	
	DIRECT	INDIRECT	DIRECT	INDIRECT	DIRECT HOURS	INDIRECT HOURS
PREPARE DIE PLATE	-	-	-	-	-	-
FABRICATE ELECTRODES	16	-	0	-	16	
MOUNT ELECTRODES AND SET-UP	4	-	1	-	3	
PREPARE N/C TAPE	-	0	-	4	0	(4)
MONITOR EDM EROSION:						
ROUGH	4	-	0	-	4	
FINISH	4	-	0	-	4	
TOTALS (STRAIGHT TIME)	28	0	(1)	(4)	27	(4)

N/C EDM RUN TIME (HOURS)------------8

SYSTEM COST (APPROX.)-------------$90-110K

REPRESENTATIVE COST COMPARISION:

CONVENTIONAL FABRICATION VS N/C EDM

(DATA IS BASED ON FIG. 2 TUNGSTEN CARBIDE REPRESS DIE)

FIGURE 3

piece in a condition that will produce the required squareness, parallelism, and locations of the part features with respect to the machine crosslide-axes and wire transport axis.

Now, you should consider these three requirements with sufficient detail for clarity of the concepts that are involved. First, we must assume that the machine is available. This is a good assumption because existing setups can be easily broken and reestablished if a super hot requirement occurs. It is relatively easy to make the machine available, if the requirements occur.

Next, the method of tape generation for a complex contour of the type we are to erode will vary depending on tolerances on the part produced.

If part tolerances are in the ± 0.003-inch and greater range, data for describing the profile on tape could be obtained from a 10 or 20X size layout, a precision protractor with ± 5 minutes of arc resolution, and a precision scale graduated in 0.01-inch increments. It will be necessary to measure included angles of arc segments, arc center locations, lengths of straight-line segments, and angular positions of straight lines in space, etc. These data will then be punched into a control tape to run the machine tool. Time for our job should be less than 4 hours, if an experienced programmer handles the job.

If tolerances on the part produced from our repress die were less than ± 0.003-inch, computer programming techniques would probably be required to describe accurately the part profile and supply data for the N/C tape. Turn around times would vary widely; however, a time of less than 4 hours would be a realistic consideration. The job could be completed in much less time if the installation is up-to-date as far as state-of-the-art equipment.

Our third requirement, job setup, requires no more-or-less than a good precision milling setup. Wire electrode and critical work surfaces must be properly aligned (square and parallel) with respect to the work plane. Special tools that may be required are usually provided by the tool manufacturers to facilitate the setup operations.

After the tape is ready, the workpiece is setup (usually accomplished in less than 1 hour) and the power supply settings are calculated. Then the controller buttons are depressed and N/C and EDM, with all of their advantages, takes over. With the job started and running satisfactorily, the operator will return to some other task and probably will not bother with the N/C EDM job for several hours. Operator intervention is required only for such tasks as securing the die dropout piece, emptying wire electrode takeup spool, and other similar tasks. The EDM will take care of removing workpiece material while N/C does the driving and stopping operations.

Considering average capability in the shop, the job is begun and running in 5 hours or less. A punch or die plate erosion operation may run for 20-30 hours; the direct labor requirements for fabrication of the complex contour would usually be limited to 1 hour or less. You run N/C EDM in-cycle to other machining processes. Leave the machine alone and it will continue to run until a program stop or "end of tape" command is read into the controller.

When the die has been completed, the same tape (or a minor modification of it) may be used to erode punch, stripper plate, and other parts. Controller functions are used to provide proper offsetting of the wire electrode to generate precise relative clearances between the die set components. Tight process control can routinely produce clearance of 0.0001-inch or less per side.

Usually, the accuracy of the tool produced will be in the low tenths (0.0001-inch) range from the programmed values if the set-up is correct, if proper electrode offset values are entered correctly into the controller, and the process variables; held to the desired values.

Your total labor requirements for the N/C EDM method would be 4 hours indirect (programming) time and 1 hour direct (set-up) time. Conventional fabrication, using the most practical available method EDM) will require 24-30 hours of direct labor for fabricating only the die plate opening. One-half to two-thirds of this time would be used in making the EDM electrodes. The remainder would be required as dedicated direct labor for monitoring the EDM operation, electrode replacement, tool resetting, and other operations.

While secondary operations for producing the rough die plate may vary for conventional and N/C EDM methods, it is usually safe to assume that fabrication times that satisfy the N/C EDM set-up requirements would be less than time requirements for the conventional method. In addition, you might also save some costs related to preparation of the workpiece for the N/C EDM operation.

Pressworking tools are excellent examples of ideal workload for N/C EDM. The making of these tools is highly desirable because fabrication by alternate methods would require significant labor costs.

But, the flexibility of the N/C EDM process makes it adaptable to other machining areas as well with similar outstanding savings potential.

Assume, for example, that you require 50-100 small parts (part profile must fit into available N/C work plane) of complex profile with a tolerance ± 0.003-inch on most dimensions. Figure 4 illustrates a good example. If, you decided that a punch and die set is impractical for this quantity, (the probability of follow-up orders might justify a die set) the parts could easily be contoured from sheet stock in batches of 2-20 pieces, depending on part thickness. Because each load would be run in-cycle to other machining operations, an operator will be doing other operations, (drilling, tapping, etc.) on the same part while N/C EDM accomplishes the profiling. Profiling operations are extremely accurate and repeatable and the eroded surfaces can be used for location reference in subsequent operations.

One-time labor costs would include set-up and tape preparation. Other operator time charged against the job will be usually limited to loading and unloading of parts and minor maintenance. Any loading and unloading sequence should be achieved in less than 0.1-hour per load. When a load is 10 parts, direct labor per part charged against the N/C EDM operation falls into the 0.01-hour or less range, a respectable charge when evaluated by any criteria. This point is worth remembering: Once a job is set up and running satisfactorily, the direct labor required to produce parts is generally insignificant. In most cases, the cost is insignificant to the point of permitting an N/C EDM operation to be run in-cycle to other N/C or conventional operations. Usually, more time is required to ring on and off a N/C EDM cycle than to perform the load/unload sequence!

In these two examples the requirements are completely different from each other. In the first case, a single copy of a special tool is to be produced for pressing P/M parts. The second example involves actual contouring of part profiles from sheet stock. In the first case, we require precise control on punch-and-die plate clearances; while in the second case, we need a capability to contour a relatively high precision (low tolerance) profile in a part or stack of parts simultaneously.

In both cases, we need sharp corners and fillets, good surface finish and relatively fast speed. We have all these capabilities with N/C EDM. Also, in both cases, most of the direct labor was eliminated by taking advantage of N/C and EDM capabilities. Precision contoured parts and tools are routine products of the process. Differences in the types of work which may be handled are not a problem. Flexibility is unparalleled.

Routine N/C EDM applications will include the two examples discussed above; pressworking tools and small job lot contouring, and a greatly diversified collection of other types of work. Figure 5 lists good work load candidates. Machining

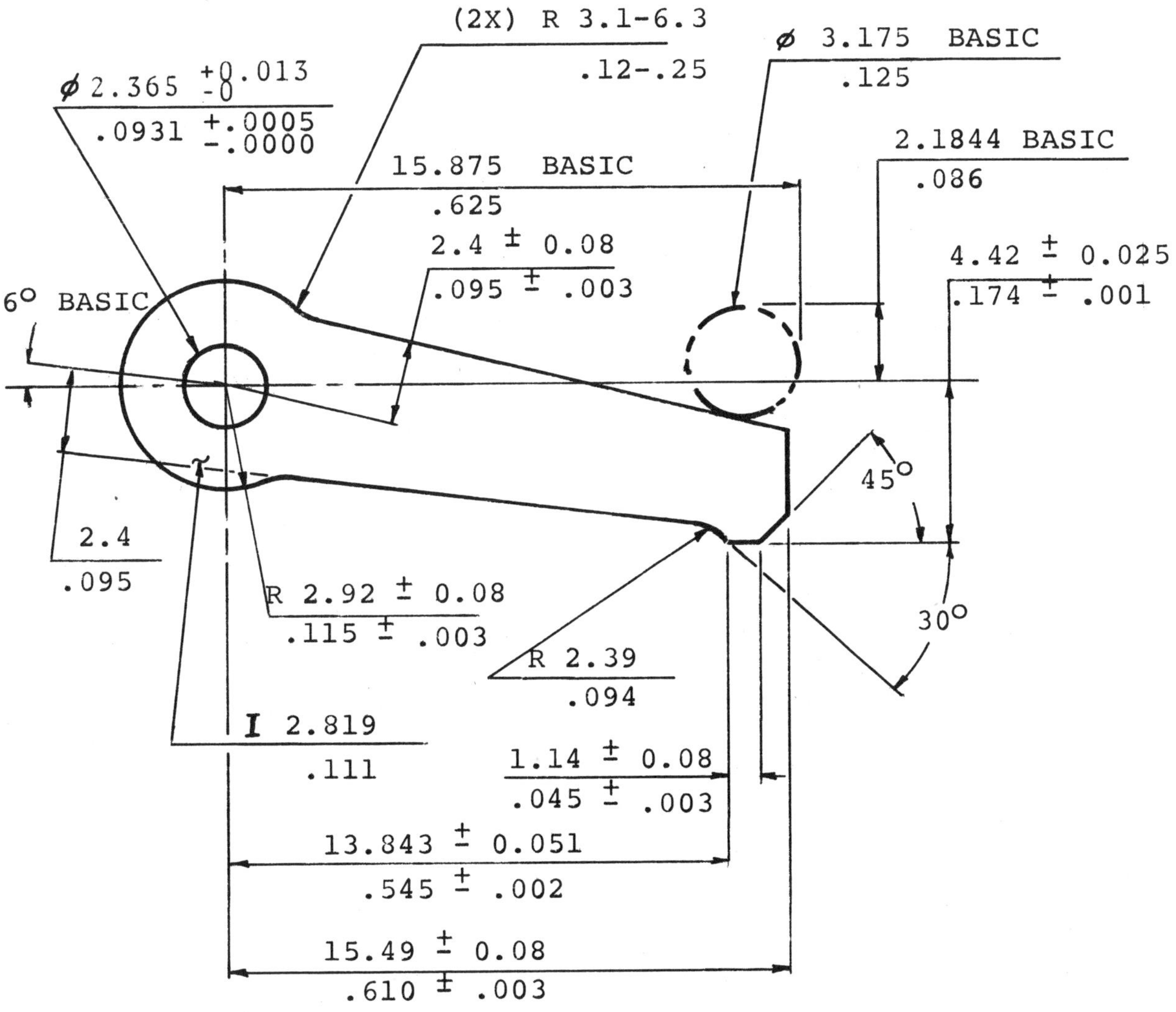

(2X) R 3.1-6.3
.12-.25

ø 3.175 BASIC
.125

ø 2.365 +0.013 -0
.0931 +.0005 -.0000

15.875 BASIC
.625

2.1844 BASIC
.086

2.4 ± 0.08
.095 ± .003

4.42 ± 0.025
.174 ± .001

6° BASIC

45°

2.4
.095

R 2.92 ± 0.08
.115 ± .003

30°

R 2.39
.094

I 2.819
.111

1.14 ± 0.08
.045 ± .003

13.843 ± 0.051
.545 ± .002

15.49 ± 0.08
.610 ± .003

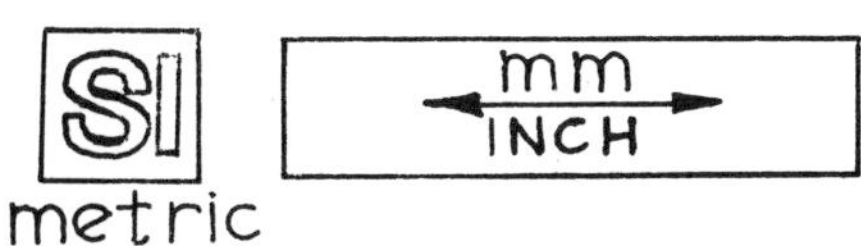

FIG. 4

SMALL PART FOR PROFILE CONTOURING

PRESS WORKING TOOLS

SMALL PART CONTOURING
- ENGINEERING PROTOTYPES
- LOW QUANTITY PRODUCTION RUNS

TOOL PROFILES
- TEMPLATES
- NESTS

NARROW SLOT MACHINING

SINGLE CRYSTAL ERODING
- SLICING
- CONTOURING

ROUTING OPERATIONS
- SQUARE INTERNAL CORNERS
- FRAGILE PART CONTOURING
- HIGH PRECISION PROFILING

COMPLEX RADIAL SHAPE PROFILING
- GEARS
- PULLEYS
- RATCHET WHEELS

'HARD' MATERIAL PROFILING
- TOOL COMPONENTS ERODED IN HARDENED CONDITION
- TUNGSTEN CARBIDE

FIGURE 5

GOOD N/C EDM APPLICATIONS

of hardened steel and tungsten-carbide guide bushings is a good application. Contours in tools or parts calling for square internal corners, (less than 0.005-inch R) are readily machined.

Also, consider machining complex profiles in small parts as specified in a routing operation. The parts may be small with thin sections.

"Impossible" jobs such as square internal corners, narrow slot machining (less than 0.005-inch) in tungsten carbide or hardened steel, single crystal rod machining, erosion of conventional EDM electrodes, with allowance for roughing and finishing generated from the same N/C tape, are typical applications. Square internal fillets can be cut in female electrode material for conventional EDM. As with other N/C operations, one tape may be used to produce as many copies as needed. This advantage makes it feasible to produce spares (punches, die blocks, conventional EDM electrodes, templates, etc.) at costs that will allow you to compile a significant return on your machine tool investment. When a job quote is based on a specific order size, <u>any</u> reruns of the same job should be highly profitable because the only actual costs associated with the job would be material, machine time, minor secondary operations, and set-up.

The case for N/C EDM thus far has shown significant savings in direct labor for production of parts and tools. So what is left for an encore? Plenty!

With negligible machining forces, fixturing requirements do not have to address cutting forces. Locating, ease of use and set-up, load-and-unload speed, and other factors become the design criteria. Reduced fixture size and complexity result in low-cost tooling, free of frills and design expense. In many cases, you can make tooling from the same control tape used previously to contour part profiles. A good example of this application is a part "nest". Another of the tooling advantages is the fact that a wire electrode may be used to pick up part features (bores, edges, etc.), and much of the necessary "tooling" capability is built directly into the machine tool. The wire electrode may be located within ± 0.0001-inch of the features.

Quoting N/C EDM jobs for profitable return is somewhat of a personal decision. Some shops are preparing quotations by computing costs at conventional rates and labor content and deducting a fixed percentage (5-10%) to make the bid more attractive. Based on your punch-and-die example, savings to the customer may be 5-to-10 hours on a 100-hour job. Profit will be based on the ability of the shop to program the N/C EDM system efficiently; however, a <u>50-hour</u> profit (dollar equivalent on a job as described in the example will be typical.

As more shops use N/C EDM, this method of quoting jobs will probably change to the extent that the percentage deducted from the estimated hours--based on the conventional approach--will increase. That remaining percentage, covering costs and profit, will be determined by programming efficiency and other costs associated with N/C tape production. This assumes equal competence in the capabilities of their direct-labor competition. The shop with superior tape-generating procedures will be in the best competitive position.

Attention to the problem of programming should be a major effort before N/C EDM is installed. The best approach is a reasonable study of the programming support requirements before a machine tool purchase order is placed. The study should accomplish two purposes. First, the parameters for the required precision must be established for the type of workload to be handled. Second, as a result of these parameters it will be possible to make an accurate economic justification, based on the required range of precision of work to be handled, because these factors will directly affect programming costs.

The cost of programming must be considered in the justification. If pressworking tools for processing parts in your plant are to be your primary product, with a typical ± 0.01 - 0.02-inch tolerance, most programming data will be obtained satisfactorily from a multisize layout of the part profile. Precision punch and die-and-stripper plate clearances are obtained by varying electrode offsets on the machine tool control. However, if your workload mix is mainly profiles of tools or parts with tolerances in the low tenths range (0.0001-inch), manual data recovery will not be practical. Some level of computer-aided programming will be probably necessary for realizing competitive programming cost within reasonable turn-a-round time.

The scope of computer-assisted capability available today ranges from simple trigonometric and arithmetic processing using relatively simple stored algorithims, through sophisticated interactive techniques. In the more sophisticated systems, you may actually describe a part geometrically, store the geometric profile and interrogate the computer for necessary program data. Once these data are developed, execution of an output sequence will generate a tape and a hard copy printout. Some computer installations use error reduction or minimization techniques; others use programs to index spatially a single gear tooth to produce a computer-stored description of a complete gear or timing pully. The possibilities are unlimited; however, you obtain only what you are willing to put into the system and that fact should be recognized when you are considering your cost of programming and your actual savings.

With the great latitude and depth of computing that is available in the software and hardware market today, you will not have a problem in obtaining the competence level that is required in software and hardware support. If you have access to a time-sharing or onsite computing system, special computer programs may be developed by your own personnel. The ideal profit condition will be attained only if your requirements match your available programming and hardware levels. Do not overbuy or overestimate your requirements for computer support. If you exceed your requirements, it will initially present a weaker case for justification of N/C EDM and ultimately reduce your apparent return on investment and increase the payback period. You can always add to your capability at a later date if the workload justifies the added capability.

A manufacturing process without inherent faults or limitations is unbelievable and unavailable. N/C EDM comes close to being faultless; however, there are some facts of life which must be remembered.

As with conventional EDM, the effective size of the electrode--the finished electrode size plus overcut allowance--is dependent on the material being eroded, eletrode material, flushing, power levels, and other process variables. Therefore, effective electrode size must be determined before committing an electrode wire to the workpiece. For precision-tolerance work, optimum results may be obtained by running a "test cut" in the workpiece material to determine overcut, surface finish, and other conditions which are process-variable dependent. Running the actual job with the required electrode offset and the same process parameters will produce the desired results and precision. Because the test cutting will require machine time which is nonproductive, this testing should be classified as a process disadvantage.

For those cases where tolerances on parts are not as precise, the test-cut route should be used only until enough experience is gained to provide a data base for overcut prediction. A prediction of overcut, based on past experience, will usually fall into the 0.0005-inch range from the actual overcut. Here again, workload precision will determine the required method.

Another of the potential problems inherent in the N/C EDM process is the apparently slow rate of material removal. However, a comparison of the actual time required to move a job out the door, or into stock, will usually show N/C EDM on top over higher-labor conventional methods of fabrication. For example, considering precision profiling work in a hardened steel or tungsten carbide die plate, an experienced operator using grinding techniques will certainly remove more material per unit time if the period of measurement is confined to the actual grinding operation. If, the total time charged against the job is examined, the nonproductive aspects of conventional fabrication will appear. For example, in any contour grinding operation, wheel redressing, resetting, and changing are costly requirements for producing high precision work. If the workpiece is made of tungsten carbide, and diamond grinding wheels are required, obviously wheel cost is also a factor to be considered for conventional fabrication.

The cost of electrode wire for a week of N/C eroding of carbide is approximately $15.00. Compare this expense with the cost of the diamond wheel consumed for the same quantity of material removed by jig or surface grinding. The savings are substantial.

A representative completion time for running a job with conventional techniques should include redressing, resetting and wheel changes. N/C EDM run time will usually require less actual time, in shop hours, to complete the same basic profiling. The cost of conventional fabrication must include labor for redressing, resetting and wheel change plus wheel and other tool costs. Add to this cost the normal nonproductive factors inherent in any operation requiring operator attention. It is very likely that total job cost will far exceed N/C EDM costs.

If we stretch a point, the need for comprehensive operator training might be considered as a disadvantage. N/C EDM will not be producing at its maximum cost-saving potential unless the operator is proficient in handling a wide variety of workload situations without excessive "hand-holding" from indirect manufacturing support areas. It does not make good economic sense to assign just any "warm body" to run an N/C EDM system. The operator must be highly experienced in making precision setups, preferably of the light, vertical or two-axis N/C milling equipment and a capability to understand the implications of changes to power supply parameters, controller functions, and effects of electrode offset. It will require time, perhaps a man-month or more and a couple of scrap jobs to attain the desired level of competence.

If your shop cannot provide this type of skilled operator or the time to train a capable individual, then you will definitely be trying to run N/C EDM at a disadvantage. Both your profit and your regard for the process will be lowered.

There may be other disadvantages of the process, which are peculiar to a specific shop environment; however, these will be uncovered while you are examining your individual requirements.

Looking at the other side of the situation, another important advantage of N/C EDM, which is especially significant when a high pressure job situation arises, is the minimization of operator error. Once the job is set up and running, the pressure is off the operator. If the N/C tape, setup and operating parameters are acceptable, results will be acceptable and insensitive to the constant job queries and other error-producing factors which might influence a machine tool operator resulting in a costly and untimely error.

A discussion of the disadvantages of N/C EDM usually produces some additional advantages as outlined previously. Obviously, the multitude of advantages offered by the process will vary with respect to individual shop capabilities and requirements. The strong point to consider is the fact that overall costs for moving a job through the shop and out the door, are reduced with the use of N/C EDM which offers a clear cost advantage for high labor content tasks.

While N/C EDM has had a relatively unheralded beginning, the many advantages of the process will surely catapult its popularity to highly respectable levels in the immediate future. The early suspicions and caution have been pushed aside by user success. Process versatility will allow adaptation to many shop requirements, some of which are yet to be developed. A shop which lacks N/C EDM capability will not be in a position to compete with a shop possessing the capability if all other shop abilities are equal.

One final point deserves mention. At a time when skill shortages of tool, die-and-model makers is approaching a critical point; the advent of N/C EDM could not be more timely. While the process will *not replace* highly-skilled machining specialists, it *will enhance* their capabilities and increase productive output. Fewer men will be able to handle a larger workload.

Numerical Control has had a significant impact on production costs and philosophy. The latest brain-child of the N/C revolution, N/C EDM, combines the necessary control features of N/C with the remarkable machining capability of EDM to produce a process which yields substantial savings in areas where high skill and great cost were previously required. When you can remove high skill and high cost out of a job during routine operations, that's progress. N/C EDM offers this progress.

Reprinted from: *Modern Machine Shop, February 1976*

Traveling-Wire EDM Carves Solid Carbide Punches and Dies That Match

The segmented carbide die is a thing of the past for one stamper who cuts from the solid with CNC traveling-wire EDM. The solid dies have much greater life due to extremely accurate clearance and location.

By KEN GETTELMAN, Associate Editor

The table doesn't seem to move at all. The only evidence, other than sparks in the gap, of any cutting action is the finely profiled kerf left by the traveling-wire electrode directed by CNC (computer numerical control). Traveling at an average rate of 1½ inches per hour in one-inch-thick steel, or ¾ inch per hour in carbide to machine a ten- or fifteen-inch profile is not going to break any speed records. Yet, Allied Pacific Metal Stamping, Division of Tower Industries, Inc., of Santa Fe Springs, California, has reaped a host of benefits in the shift from the traditional to the electrical discharge machining method of shaping complex die sets. These blanking die sets, which include both the punch and the mating die, are made from carbide and high speed steel.

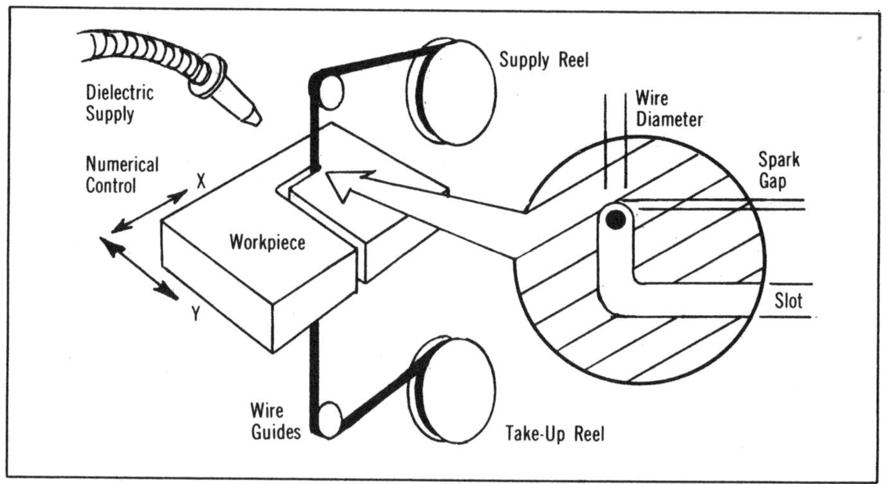

The electrode is a wire which travels from the supply reel through the workpiece to the take-up reel. The profile of the cut is determined by NC command to the X and Y axes of the table.

Foremost among the benefits is a control over position and clearance not obtainable by conventional means. If it is desired to have a 0.003-inch-clearance between punch and die on all faces, ends, radii and slopes, such a factor is programmed into the computer control. The finished product comes out exactly as programmed and the dies last much longer.

Secondly, the laborious task of sectioning dies and fitting them together has been eliminated. It is now possible to work with solid dies—even those with complex contours.

Finally, the whole die-making process is much faster since grinding and matching have been eliminated. Die setup time is also reduced.

EDM has been a viable metalworking technique for at least 30 years and it has been used in the United States for the past 25 years. Even the traveling-wire variety has been around in one form or another for the better part of ten years, but it is just now beginning to win general acceptance.

For the record, EDM is a spark-erosion process and should not be confused with ECM (electrochemical machining), which is essentially a deplating process. Most EDM machines in use today employ a solid electrode, often graphite, that plunges into the workpiece and slowly advances as sparks jump across the gap between electrode and workpiece to erode away the workpiece material and leave a shape that is a mirror image of the electrode. The spark actually is forced through a dielectric insulating fluid, often nothing more than a thin oil, to impinge upon the workpiece. The dielectric fluid is pumped through the gap between electrode and workpiece (often the gap is only 0.002 to 0.003 inch) to flush away the detritus.

The traveling-wire electrode operates in a much different manner. Instead of a plunge cutter, its functioning is much like that of a band-saw blade. It literally cuts a path through the material. Like the band saw, the traveling-wire EDM machine moves the workpiece past the wire electrode, but since the cutting rates are very low, the movement is usually done with a numerical control unit. In fact, the machine may run an entire night totally unattended.

The unit at Allied Pacific Metal Stampings is a Charmilles Model F40 that will mount a 20″ x 20″

workpiece and has an X and Y table movement of 10″ x 10″ under full computer numerical control. The investment for the machine and control unit with the necessary peripherals came to about $175,000. It is earning a very substantial return on investment with its unique capabilities in terms of dies made for outside customers and those created for use within the Allied plant, which specializes in precision metal stampings.

Drawing To Die

The process from workpiece drawing to finished die is similar to that for most NC machine tools. It starts with Steve Andersen, who both programs and runs the TW/EDM. He has developed the insights and is fully attuned to the potentials of the combined CNC and TW/EDM resources. In short, he has mastered many "tricks of the trade," which add up to a productive and creative use of the investment.

The initial programming step follows the universal NC pattern. Mr. Andersen studies the workpiece drawing and decides how best to approach the machining of a particular punch or die. He determines how he wants the workpiece loaded on the machine and where he wants to start his cut. This is most important since, as with a band-saw blade, an inside cut must start with a hole in the workpiece. Once determining the general approach, he then literally talks the wire around the profile of the cut. The talking is done with a series of statements and definitions first written down on paper. Those statements become the workpiece document. A punch and die of reasonable complexity may require from 20 to 30 statements in the workpiece program documentation. The average preparation time may run from 10 to 30 minutes.

Mr. Andersen then types out his program statements on a teletype unit for both proof reading and tape punching. It is important to understand that the tape at this point does not contain the hundreds or thousands of X-Y coordinate points needed to EDM the workpiece. It merely contains a series of statements that describe the geometry of the workpiece in a format that can be acted upon by the computer element in the CNC machine control unit. Thus, the actual tape may be only two to four feet in length.

Once in the CNC, the instructions are "read" and the computer then proceeds to generate the actual coordinate table movements necessary to machine the workpiece. The system has been so successful that Mr. Andersen makes no documentation or permanent tape record of the actual machine movements. He keeps only the original source document, which is merely a set of workpiece geometry definitions for the computer to work on.

Here is where the CNC has earned its way. If it were necessary to manually calculate all the coordinate points for a workpiece, it would take days or weeks for even the best programmer. If it were done off-line on an outside computer, the resultant workpiece tape with all its thousands of coordinate points would be hundreds of feet in length. By having a dedicated computer in the control unit do the work, the input is brief, the tape reading is even more brief and the problems of extensive tape reading are eliminated. The computer calculates the necessary coordinate points and then communicates very well with the machine electrical servo drives. It also figures the necessary offsets for wire diameter and the desired punch and die clearance. None of this is committed to "hard" tape copy.

The computer is a Digital Equipment Corporation PDP-11/05. For the TW/EDM purposes, it has been equipped with a PROFIL 1A processor language. It is capable of interpolating two-axis straight line, circle and arc cuts. If the profile of the workpiece involves more complex curves, Mr. Andersen works out his own techniques of breaking them up into a series of arcs and minute straight-line segments to always have a tangential blend and a very close approximation of the real curve definition.

Since Allied produces a number of gear blanking dies, there is also a gear processor language. By defining the pitch diameter, number of teeth, type of tooth profile and other important parameters, the computer will calculate and then execute the machine movements, along with proper offsets, to generate a mating gear punch and die.

The computer also has other capabilities. It will store macros or routines that are repeated on a frequent basis so that it is only necessary to calculate them once and then call them up for use with a single call statement. The computer will also calculate appropriate machine movements in either inch or metric—it could care less.

Offsets are critically important and it is here that the computer really comes into its own. The traveling-wire electrode is exactly 0.185 millimeter (0.00728 inch) in diameter. The erosion overcut is very small and very predictable—so much so that from a single program giving the nominal outline of either the punch or die and the desired clearances between the two, the computer will calculate the necessary offsets. Workpieces are produced to a ±0.0005-inch tolerance as a matter of course, to ±0.0002 inch in special instances and to ±0.0001 inch in unique applications if willing to spend the time in a trial-and-error phase. These tolerances will be uniform around the punch and die—not just at certain check points.

Mr. Andersen usually programs a couple of reference or dowel-pin holes into his punch and die sets to assure easy and proper mounting once the components are out of the EDM unit. All components in a set are generated by the same NC program and the reference holes will assure that the same relationship extends to the actual die mounting and use.

In essence, the computer becomes the instrument that takes a limited amount of human input in the form of workpiece definitions and instructions and converts those statements to the multitudinous outputs for the TW/EDM machine to actually profile them.

The power supply for the TW/EDM machine differs extensively from that found on most control units. The power settings are very limited. There is only one electrode material that will ever be used—the 0.185-mm brass wire—and only one dielectric selection. It so happens that experience has proved that deionized water is the best dielectric for traveling-wire use. Each spark generated by the power supply is actually a full wave that goes through both a high positive and a high negative peak on each cycle with an average zero voltage. This prevents any electrolytic or plating action from taking place between the brass wire and the carbide or steel workpiece. With the extremely thin wire, the current output is only six amperes and the frequency is not a selected number of cycles per second as would be the situation with most conventional EDM power supplies.

For TW/EDM work, there is a variable frequency which consists of short bursts followed by longer pauses. All of it is controlled by the power supply which, in turn, is acting upon the conditions actually taking place in the cut. The reason for the totally different power-supply design is the necessity for assuring that the walls of the cut are not tapered, bell mouthed, bowed, or machined with some other equally undesirable condition. It is also necessary to have a good finish and Mr. Andersen reports that he secures a 30-rms reading with no difficulty. On occasion, a machined surface will be smoothed with a "kissing" grind or with light abrasive blasting. No dimensions are changed—just a refinement of the workpiece surface.

Proveout

With machining rates of only ½ to 2½ inches per hour, some means of NC program checkout is desirable. Mounted to the machine table is a platform on which a sheet of paper is taped. A holder, mounted to the wire guide-arm, contains a pen. During checkout, the speed of the machine is set to "full" and is totally independent of cut conditions. A tracing of the cut is made in a few minutes. Again the computer comes into play. It is possible to scale up or scale down the workpiece or any portion of it. Once it has been determined that a workpiece program is correct, the workpiece itself is mounted, the machine is zero referenced and the cutting action is initiated. The machine is theoretically capable of running the next 40 to 50 hours unattended if the workpiece program is that long.

Cutting Tool

The cutting tool, which is the brass wire, never comes in direct contact with the workpiece. Electrical sparks that actually do the workpiece eroding simply jump the gap between wire and workpiece. If the wire were not constantly traveling from a payout reel past the workpiece to a takeup reel, there would be enough eroding action that the wire itself would soon disappear. Hence, a fresh supply is constantly drawn into the action arena.

One of the first questions usually asked of a TW/EDM user is "How much does the wire cost and how long does it last?" As Steve Andersen points out, a reel of wire will last 40 to 50 hours and the cost is approximately one dollar per hour. The cost is so low that no attempt is made to re-use the wire. Some diameter is lost in the eroding process.

Although the wire is a relatively inexpensive item, it is still a precision cutting tool. When received from the supplier, it has a nominal 0.200-mm diameter. The Charmilles EDM machine is equipped with a wire drawing die so that the diameter is reduced to a consistent 0.185-mm diameter right on the machine itself. The wire then travels through an annealing chamber mounted on the machine frame. This is one of the factors in obtaining a precision cut. Other factors include a minimum machine resolution of 0.001 mm, which is about 40 millionths of an

1

2

3

The new way to make a precision carbide punch and die: Steve Andersen studies the drawing and prepares his manuscript (1) of some thirty geometry statements. He then punches out a tape on a teletype (2) and has a printout to check for typographical errors. He enters the tape (3), which contains only the source statements of geometric data, into the minicomputer, which is an integral part of the Charmilles traveling-wire EDM CNC unit. A final check is a three-to-one tracing enlargement of the punch and die outline (4), which is produced from the data on the source tape. Thirty machine hours later, he has a punch and die for a crown nut (5) with perfectly matching form and clearance well within a plus or minus 0.0005-inch tolerance.

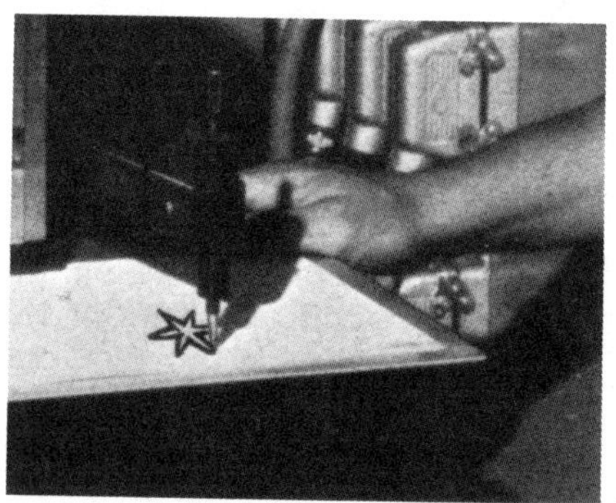
4

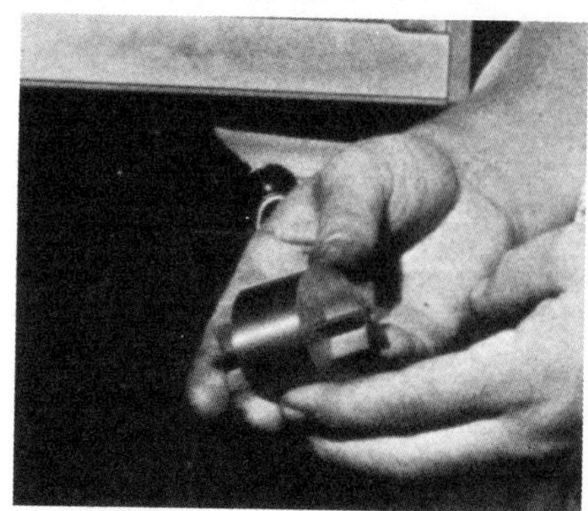
5

inch. Wire threading is no problem. The machine is actually self-threading and if it is necessary to break the wire for setup or other purposes, the rethreading is simply a matter of starting the wire and then affixing the end to the take-up spool when it appears.

Dielectric

Water is normally considered an electrical conductor. This is true if there are many different ions in the water. A completely deionized water is a very poor electrical conductor, but it does have very excellent penetrating qualities. For these reasons, deionized water is used as the dielectric fluid. It has one added advantage—it is clear and enables the observer to see the workpiece and the cut. With an overcut of less than 0.001 inch, it is highly important for the dielectric to be drawn down into the cut area. Deionized water will do this. As the eroding action takes place the resulting detritus does tend to contaminate the water, but the recirculating system and filters are designed to filter the water as it comes out of the cut.

Die Taper

In most blanking dies there is a one-half to one degree taper for slug dropout. That taper can be machined in as part of the TW/EDM operation. This is done with a rotating head or "Z" axis, which is slightly offset from the perpendicular. The function of the "Z" axis computer control is that of always keeping the rotation in a plane normal to the surface of the die being machined. Thus, the taper is uniform throughout the entire die section—both straight and curved outlines.

Slug Dropout

In any traveling-wire EDM work there is the point where the cutout is complete and the slug will drop. Steve Andersen uses his own approach to that problem. Just before the dropout occurs he stops the machine and lowers the dielectric water level. He then takes a piece of flat scrap and places a drop of Eastman 910 adhesive to each end of the flat. It is then affixed with one end to the parent material and one to the slug. When the cut is completed there is no sudden shorting due to direct contact made by a falling or teetering slug. A sharp tap of a hammer will then break the adhesive bond.

Taking Advantage

The slow cutting speed is not an unmitigated drawback. Steve Andersen often plans his work so that a long cut will be started toward the end of the workday. The machine will be left on overnight and the computer control and power supply will very quietly and efficiently work on through the wee hours. In the event of an unlikely occurence such as a wire break or some other accident, there are enough sensors that the machine will shut itself down. The experience to date has been excellent and machine utilization has been high during the unattended night hours. If the program comes to an end or to a programmed stop, the unit will automatically shut down.

Hidden Factors

The quality of human operation, programming, and maintenance is always the hidden factor of NC or any other process. The human programming and processing factor can be a very great item in numerical control due to its unusual capabilities and the size of the investment. When NC works well, it works very well. If not, the results can be most disappointing.

When the TW/EDM was installed in early 1975, Mr. Andersen was given a week's training on site. It proved adequate. Since that time he has continually sought out the tricks, short cuts, ingenious programming routines and methodologies that are the hallmark of any flourishing NC application. He has an intimate working knowledge of the machine and its capabilities. They have paid off. A wide range of complex blanking dies has been produced. They have been programmed with a minimum input by knowing how to make the computer do most of the work. In short, the installation has taken human concepts and turned them into a part producing reality with finesse and efficiency. **MMS**

Reprinted from: Production Engineering, March 1977

Electronic bandsaw sparks a revolution

Wire EDM—electrical discharge machining with a thin, copper wire electrode—is moving out of the toolroom and onto the shop floor. It is proving to be an economical way to carve out prototypes and small-volume runs of hard-to-make parts.

By DONALD E. HEGLAND
Associate Editor

Machining with wire is a special variation of electrical discharge machining. As in conventional EDM, the cutting mechanism is spark erosion. Material is vaporized from the workpiece by an electrical discharge. The big difference is in the electrode—the "cutting tool." Familiar, plunge-cut EDM employs a complex electrode, shaped to either match or mirror the shape of the finished part. The electrode is sunk into the workpiece, gradually eroding it to produce the required form.

Wire EDM employs an expendable wire electrode that is guided by a numerical control system to generate the required form in the workpiece. Think of it as an N/C, omnidirectional, electronic bandsaw with one exception—the wire electrode never touches the workpiece. High-resolution servos maintain the necessary spark gap between wire and workpiece.

The wire unreels continuously from a supply spool, passes through the kerf in the workpiece, and is taken up on a discard spool; hence the other common name for the process—traveling-wire EDM. The wire only makes one trip through the workpiece and is discarded, therefore electrode erosion is insignificant.

Tool and diemakers have embraced wire EDM with a passion; some users say it's the best application for N/C in the toolroom and one says the marriage of N/C with wire EDM is the greatest single advance ever seen in toolmaking. But the benefits don't stop at the toolroom door. Clever production engineers are finding more and more uses for wire EDM on the production floor.

A few limitations—a lot of advantages

Wire EDM has a lot going for it but like anything else it isn't a panacea. Although few, the limitations need to be considered. First, your workpiece has to be electrically conductive—no form of EDM works on an insulator. Second, you can't cut blind cavities—you can only do through-hole jobs. Third, wire EDM machines cost from $80,000-200,000, depending on size, capabilities, accuracy, and system sophistication. So, to get the best return from your investment, you have to be a little cagey about how you use wire EDM. As a general rule-of-thumb, it is for the tricky and complex jobs, especially on intractable materials. Experts advise that many garden-variety jobs still belong on conventional EDM machines.

Having defined the gross boundaries of applicability, what benefits lie within? Tooling economy is a big plus. Making the complex electrodes for conventional EDM usually represents about 40-60% of the total cost of the EDM work. And the electrodes can often be used only once because they are eroded as they cut the workpiece. There are no electrodes to make for wire EDM. The only tooling cost is for the wire itself, a mere

10-20¢ per hour of operation.

Wire EDM leads to simpler die designs and slashes toolmaking costs from 50-70%. Split and sectionalized dies that require subsequent form grinding are things of the past; wire EDM cuts them in one piece. Punches, die blocks and shoes, ejectors, and stripper plates can often be cut from the same tape. And it is also being used increasingly as the most economical method of making the complex electrodes used in conventional EDM.

Even if wire EDM is only used in the toolroom, the benefits spill over onto the production floor. Steel alloy tooling can be made after the blanks are heat-treated and stress-relieved (as in conventional EDM), resulting in close-tolerance tools that don't require finish grinding to remove distortions introduced by heat-treating. Long-wearing carbide tools that might be impractical or impossible to make by conventional methods because of their complexity are easy with wire EDM. The result is simpler, more robust tools that stay in the presses longer, making better parts faster.

Quick turnaround on die maintenance or repair is another plus. The workpieces are "stored" on tape and are available on demand, so replacing a broken or damaged tool with an exact duplicate is simply a matter of loading a blank into the machine and starting the tape through the tape reader.

Cams, gears, master gages, and any hard-to-define geometric shapes—especially in the exotic, space-age materials—are practically a piece of cake for wire EDM. Workpiece hardness is not a limitation; Alnico, Inconel, Waspaloy, and moly-rhenium are just a few of the "nasty" materials that wire EDM cuts easily. And there are even advantages to using wire EDM on softer materials like brass; zero-force machining leaves no burrs and no residual stress or distortion. Accuracy and surface finish are usually poor on coarse-grain graphite cut with wire EDM and the process is not generally recommended for this material.

Prototypes and short-run parts are also ideal candidates for wire EDM. Difficult blanks can be cut in comparatively short times, and layers of sheet metal can be stacked to cut a number of parts in one pass. Making samples in the development stage means design savings too. Prototypes can often be made for half, or less, the cost of production tooling. And the pro-

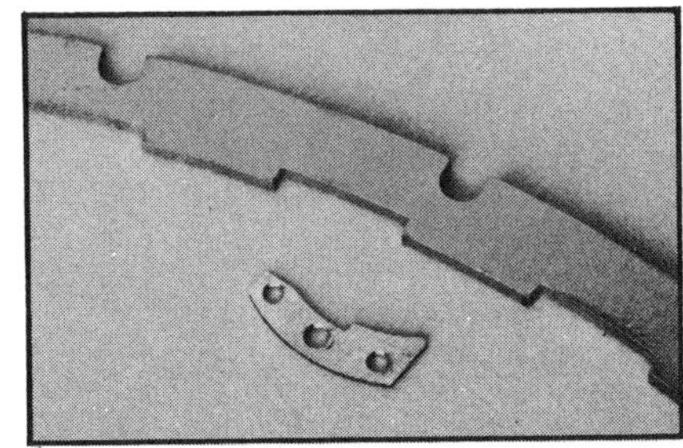

Stacking blanks makes it easy to cut quantities of thin prototype parts with wire EDM. Fifty of the small parts (0.02-in. thick) are cut in one pass in 1.5 hr. The large parts (0.08-in. thick rings) are stacked 12-high and cut in 50 hr.

Courtesy, Elox Div., Colt Industries

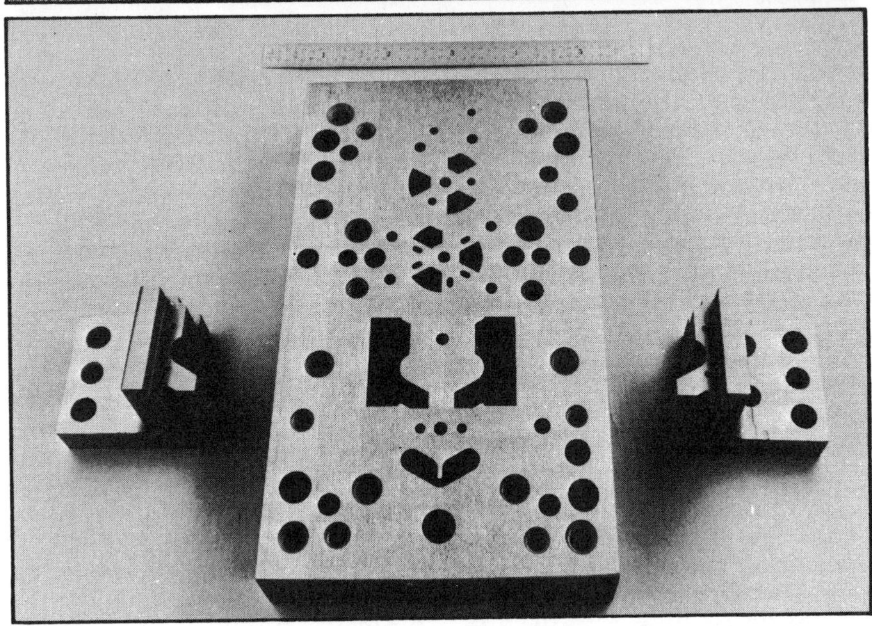

Total time to make the punch and die set (below) was 33.9 hr, including 10.5 hr for programming and 23.4 hr for cutting. Material is D2 hardened tool steel. Compacting tool set (above) was made in 31.75 hr, including 25.5 hr to cut the carbide parts, 5.25 hr to cut the brass laps and one hour for programming. *Courtesy, Andrew Engineering*

Courtesy, Elox Div., Colt Industries

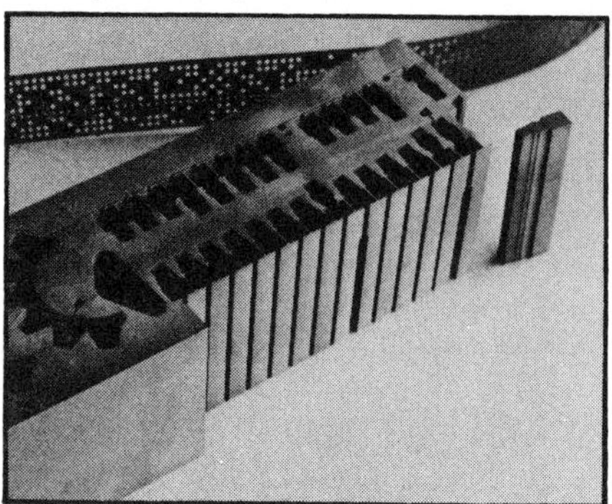

Courtesy, Andrew Engineering

Two examples of linking programs together to cut a number of punches in a continuous run. The finished punches are separated from the parent material with a single, straight cut.

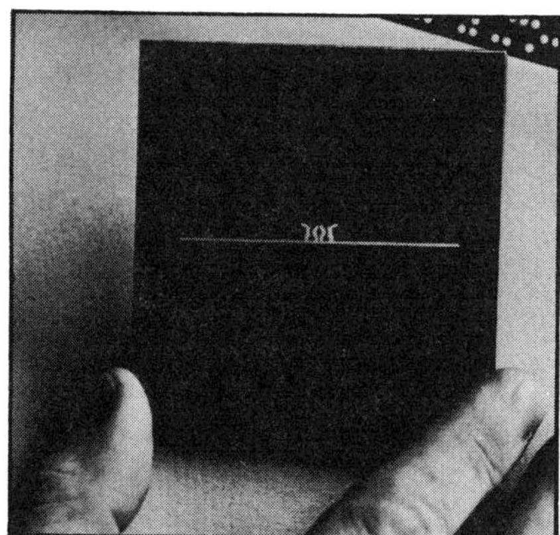

Tricky jobs are wire EDM's forte. The piece on the left has 12 dovetail slots, 0.1-in. wide by 2.25-in. long, with a radius on the bottom. Programming time—one hour. The extrusion die on the right includes 0.02-in. wide slots and 0.01-in. thick ribs with blended radii and angular straight lines. Manufacturing time—less than ten hours. Courtesy, Elox Div., Colt Industries

gramming effort isn't wasted—after proveout, the same tapes are used to cut the production tooling. Thin parts that require subsequent forming can be cut economically with wire EDM and used to establish forming station requirements at the same time that production stamping dies are being made.

A look at the process

By comparison with conventional metal-removal processes, the cutting speed of wire EDM is very slow. This is, however, deceptive. Zero lead-time for tooling, economical programming (zero time for an existing tape), and the ability to run continuously and unattended add up to make wire EDM competitive in terms of floor-to-floor time. And the linear cutting speed is independent of contour geometry, which simplifies estimating a job.

Typical cutting speeds in hardened D2 tool steel are 1-1.5 in./hr in 1-in. stock and 0.8 in./hr in 3-in. stock. The speed is not linear with thickness because thicker stock exposes more electrode area for better utilization of the available energy. Cutting speed in steel is virtually independent of alloy or hardness. Aluminum is less dense and has a lower melting point, so it cuts faster. Although carbides can be cut nearly as fast as steels, the surface finish is usually degraded because of the power required. Hence, recommended cutting speeds are from one-half to three-quarters that of steel.

The upper bound on workpiece thickness with present machines is about 3-4 in. There seems to be no practical lower bound. Material as thin as 0.002 in. can be cut with ease, either single-ply or stacked. Surface finish on parts cut with wire EDM looks like 10 rms, say users, but measures about 40 rms. The texture is matte, which is often advantageous for dies because it holds lubricant. As in conventional EDM, faster cuts—that require more input power—produce rougher finishes.

There are two schools of thought regarding the relative motion of workpiece and wire-guide system. One holds that the wire-guide system should remain stationary while the workpiece moves, as in an ordinary bandsaw. Advocates of this method claim that the wire-guide system is sturdier, providing superior accuracy. The other school favors holding the workpiece stationary and moving the wire-guide system. Those who favor this method say that controlling a constant mass—the wire-guide system—eliminates errors caused by differences in workpiece mass, and that this configuration simplifies loading and unloading. They also point out that it permits mounting several wire-guide systems on a single base. Accuracy is excellent regardless, with absolute positioning to ±0.0002 in. and repeatability to ±0.0001 in.

Water—filtered, deionized, and demineralized in a continuous closed-loop system—is the dielectric used with wire EDM; conventional EDM uses oil. Using water eliminates a potential fire hazard, hence the machine can safely run unattended. Water does a better job of flushing the debris out of the kerf than oil, and because of its higher heat capacity provides better cooling. The combination nearly eliminates the recast surface often found on parts cut with EDM.

Some wire EDM machines operate with the cutting zone—and often the entire workpiece—

completely submerged in a tank of dielectric as in conventional EDM. Other systems employ as many as three jets to direct dielectric into the cutting zone, with the runoff collected in an open tank that surrounds the workpiece. Those who favor submerging the workpiece in dielectric feel that this method provides better cooling of the electrode without agitating the dielectric. They also point out that workpiece temperature is more stable and homogeneous, and that submerging the workpiece protects it from atmospheric corrosion during machining. Proponents of the jet method of delivering dielectric to the cutting zone say that the jets keep the debris flushed out of the kerf better than submerging the workpiece does.

Kerf width is determined by wire diameter and applied power. Wire diameter typically ranges from 0.002-0.012 in., with copper used most frequently. Manufacturers generally recommend using molybdenum or tungsten to provide added mechanical strength if the job calls for wire smaller than 0.004 in. The overcut is generally 0.0015-0.002 in. per side. This can be reduced to 0.001 in. in thin materials by reducing the applied power, but cutting will be slower. Wire diameter is also affected by the minimum inside corner radius to be cut, and by the nature of the workpiece—harder and thicker materials generally call for heavier wire.

The cutting edge

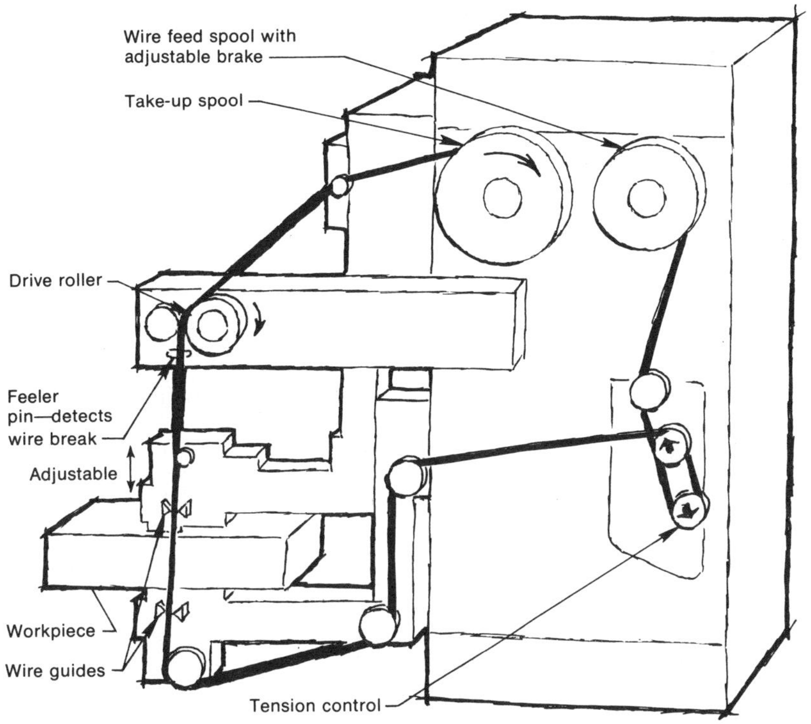

Typical wire EDM machine with stationary wire-guide system and moving workpiece. Heart of the system is the wire electrode that chews its way through the workpiece by spark erosion. Feeler pin shuts the system down if the wire breaks—a fail-safe feature for unattended operation. *Courtesy, Agietron Corp.*

N/C does it

Numerical control is the key that enables the user to avail himself of all that wire EDM has to offer; without N/C, the usefulness of the basic machine would be severely restricted. Consequently, all wire EDM machines incorporate some form of N/C. Some machines use simple, two-axis, hardwired N/C and respond either to direct keyboard input or punched tape prepared elsewhere. The more sophisticated machines are CNC, with a micro- or minicomputer built into the control system. These often include three-axis capability, useful for jobs like cutting small tapers for die relief in a single pass.

Other control systems interface with tabletop programmable calculators or in-house mainframe computers. In most control systems that include some form of computer, the computer can run the machine through one job and prepare or debug a tape for another job simultaneously.

Simple shapes can be programmed directly, using a calculator to determine coordinates from the part print. Complex jobs need computer assistance either from the built-in units (CNC) or separate computing facilities.

A series of programs can be stored in the computer and played back for continuous overnight or over-the-weekend unattended operation. The only restriction here is that all of the cuts have to be connected. Each new cut—unless it starts at the edge of the workpiece—requires manual rethreading of the wire through a starter hole. Rows of parts are often cut on a continuous basis and then separated with a single straight cut. Some users report trouble-free unattended operation for as long as 65 hr.

Information and illustrations for this article were provided by: Agietron Corp., Woodside, L.I., N.Y.; Andrew Engineering, Hopkins, Minn.; Charmilles Corp. of America, Plainview, N.Y.; and Elox Div., Colt Industries, Davidson, N.C.

CHAPTER 6

LASERS

Production Laser Hole Drilling—Now

By Albert D. Battista
President

Whilliam H. Shiner
Vice President, Laser Inc.

There are, today many laser drilling installations in the U.S.A. Many of these use Nd: Glass lasers because of the quality of the pierced hole produced, and most importantly, because the technique is cost-effective. The necessary characteristics of drilling lasers are discussed and for the attributes of high radiance, a highly-spiked output wave shape, and high reliability, the superiority of glass lasers over YAG is shown. A description of near-field imaging optics and its advantages is presented. Examples of industrial applications and the machines delivered for this work along with their costs are given.

INTRODUCTION

There is, today, a growing use of a new technology on the production-line floors of many U.S. companies - LASER DRILLING. For these companies, laser hole piercing is a major drilling method. Once believed to be a tool of the future, laser equipment has become a new tool which now has a place in industry. The easiest course for the industrial user may seem to be to ignore lasers unless no traditional drilling method works; and yet, to dismiss lasers without some investigation of the merits of this type of equipment may prove costly. Rather than dismissing lasers out of misunderstanding or fear, skeptics should be questioning how others are making use of this tool. Questions should be asked of things such as reliability, life, productivity, and economics. In other words, when can the laser be put to use drilling effectively? It is the purpose of this paper to illustrate the answers to these questions when use is made of glass laser systems. Examples of industrial production applications will follow with a brief description of the drilling machines and their costs.

HOW DOES THE LASER DRILL?

A laser is a device which produces a narrow beam of light at a specific wavelength. Basically, the laser generates a very high intensity light beam that is highly collimated with high wavelength purity. These properties allow focusing the beam to an intense spot. The resultant high power density on a workpiece from a focused laser beam, and the localized nature of the working region can melt and/or vaporize amounts of material from the workpiece without affecting adjacent material. To laser drill, a very short pulse of laser light is used. This pulse so rapidly heats the surface of the material that vaporization takes place. The resultant gases and vapor pressures generated blow away vaporized and molten material leaving a hole in the workpiece.

When all factors are considered only two solids, both ancient in the span of laser history, have so far emerged as useful production laser materials. They are neodymium-doped glass, and neodymium-doped YAG. These materials have both produced sufficient energies for hole drilling and numerious claims have been made about their relative merits. Although neither material has all the advantages desired, glass lasers have been shown to be somewhat more effective in drilling metals than any other type of laser available today, because of some inherent qualities.

There are four important parameters for laser drilling: efficiency, output pulse shape, radiance, and repetition rate. Values of these quantities may be influenced greatly by proper design or "technique" and, indeed, some of the techniques developed work equally well with either material. Comparisions of these four characteristics for both materials follows.

Efficiency

In well-designed, optimized systems, the efficiency of YAG and glass lasers are comparable. When output-vs-input curves are experimentally determined for both these types of laser systems, they are like the curve shown in Fig. 1 where the output is shown to be proportional to input above a "threshold" energy. Overall efficiencies of 2-4% are common for these systems.

Pulse Shape

By laser output pulse shape is meant the curve of output power versus time. The overall pulse shape and its duration is determined by the power supply and flashtube characteristics, and this "envelope" is determined using standard electrical pulse-shaping methods. Pulse durations of from 150 to 1500 microseconds are commonly used for drilling.

The manner in which the output occurs within the envelope is controlled by the laser cavity. Typical outputs of a Nd: Glass and a YAG laser are shown in Fig. 2. Here is shown a time trace of the output power versus time of two lasers of equal energy output. The more highly-spiked output of the glass laser is a very important characteristic when drilling because the higher peak power pulses vaporize a higher percentage of the material to be drilled and blows away molten metal leaving a hole in the workpiece that is free of recast metal. Drilling tests made with a 10 joule, 900 microsecond output from a glass and then a YAG laser on the same piece of metal showed a 50% more penetration per pulse with the glass laser's 10 joules than with the YAG's.

Radiance

The radiance of a source is defined as the radiant power emitted per unit source area per steradian, and high radiance is the reason lasers can be even used for drilling. Once again, the higher radiance from a glass laser yields a higher power density at the workpiece - another reason why they drill better.

There are two convenient ways to focus laser radiation onto the workpiece. The first method, which is often referred to as far-field pattern imaging, yields a spot at the focal plane of a focus lens as shown in Fig. 3. This technique gives good working distance ($\simeq f$) in relation to the spot size desired with small beamspread (θ) lasers. Its disadvantages are that the intensity within the spot can be quite non-uniform and also the spot size changes if the beamspread changes. As this unfortunately can occur with repetitious pulsing of the laser, this technique isn't used at LASER INC. The method which is used is called near-field imaging whereby an image is formed of an aperature placed near the end of the laser (Fig. 4). Using this method, a very uniform spot of stable dimensions is obtained. The disadvantage of near-field imaging is that to achieve a long working distance with small spot-sizes requires the laser to be put a long distance from the lens. This seems to be a small enough price to pay and accordingly, the industrial installations shown often illustrate this distance.

Repetition Rate

Certainly this characteristic of a pulsed laser system is important for drilling. Here, there is a definite advantage in the use of YAG. The very poor thermal conductivity of glass makes it much harder to cool than YAG. LASER INC. glass laser systems typically are pulsed at a rate of once per second while YAG drilling repetition rates of up to 20 pulses per second are achievable.

Reliability/Cost - Why Glass?

Although the better hole obtained because of the high radiance and highly-spiked output from a glass laser often is the only reason glass lasers are used for drilling, there are two other attributes of glass lasers - reliability and cost. Industrial users will not adopt a new technology unless it can be proven cost-effective. The power supply designs are essentially the same for glass and YAG, but the laser module designs differ markedly. A glass laser rod can be made quite long and to do so allows the use of long flashlamps. It has been well established that the maintenance cost for flashlamp replacements are by far the highest of all other maintenance and, indeed, all other costs added together don't equal the cost of lamp replacement. It has been our experience that the use of long flashlamps, with correspondingly decreased energy-input/inch, results in much higher lamp lifetimes.

The first drilling system of this type was installed in 1972 and on this and succeeding units we have been accumulating cost data as it becomes available. In the short experience that we have to date, a cost of 0.04¢ per shot or less is a good estimate for the total cost of operation (except for the cost of the operator).

APPLICATIONS - SMALL HOLE DRILLING

In truth, all the holes drilled by LASER INC. equipment are small, but this paper is especially devoted to production applications requiring holes of a diameter <0.020" and, accordingly, all comments and examples are for holes of this size.

Brass Caps

Fig. 5 shows brass parts with a single hole laser pierced to form the calibration orifice used as part of gas tank pressure-reducers. Although a variety of hole sizes are used, a common one used as an example is 0.010" diameter. These holes are produced with a diameter tolerance of ±0.0003" at a rate of 50 parts per minute. In addition to the cost-effectiveness of laser-piercing, a further savings was realized by the customer. This part was formerly machined with a flat surface to minimize drill breakage. The part is now produced by a punch-extruder at a fraction of its former cost. The laser system for these parts is shown in Fig. 6. The covers which normally enclose the system have been removed for this photograph. Here, the long path length used to give a long working distance is readily apparent. In operation, the brass caps are dumped into a hopper. A vibratory bowl, as required, takes the parts and correctly orients them on an inclined trackway. Spring-loaded jaws feed the parts to the drilling station. After the laser "fires", the part is ejected and a new one is positioned for piercing. This machine was priced at $31,000 and paid for itself in 8 months use.

Stainless Steel Cups

A similar application to the previous one is shown in Fig. 7. These are stainless parts which are pierced leaving a hole which is 0.0043" ±0.0002" diameter. The machine is similar to that shown in Fig. 6. Parts are produced at a rate of 40 per minute and at least 95% are within the required tolerance range. The parts are used as part of the spark-advance mechanism in a General Motors car. Formerly produced by an electrical-discharge machine (EDM) process, the laser machine produces the parts at 1/3 their previous cost.

Fuel-Pump Valves

A division of a U.S. car manufacturer is producing fuel pump parts for its passenger cars with the help of a LASER INC. drilling system. The laser has been incorporated into a multi-station assembly, hole drilling, and testing machine and is producing bleeder holes in cold rolled steel fuel-pump valves (Fig. 8) at a rate of 1 hole per second. The holes are 0.008" ±0.0003" in diameter and the thickness of the steel is 0.022". The bleeder holes produced by this machine assist the fuel pump action and permit improved emission characteristics.

Formerly for a hole of this size a conventional drill, mechanical punching operation, or EDM drilling would have been used. However, with the very tiny drill bits or punches used for a mechanical operation, there would have been a lot of tool breakage and wasted expense. EDM requires holding the electrode close to the work and is a much slower process than the laser system. EDM is also more expensive.

The laser was installed in June 1975 and has been used, on a 2-shift per day basis since. Except for routine replacement of its cooling system filters, it has yet to require any maintenance. The system was priced at $21,000.

Transpiration Hole Drilling

Today's high performance aircraft use high thrust engines which operate at a very high turbine inlet gas temperature. These high temperatures are made possible by new alloys and by air cooling the components of the engine. The designs require a large number of small holes ranging from 0.007" to 0.040" diameter and many are drilled at acute angles. Because of the small hole sizes, the entrance angles, and the hardness of the material, these holes can be drilled by electrochemical, EDM, or laser processes only. For much of this hole drilling work, lasers are by far the most cost effective drilling method. However, in this application, all holes must meet very rigid specifications with no allowance for error. To meet these requirements, a special laser system has been developed by LASER INC. This system makes use of two standard Model 11E lasers operated-in tandem to deliver the required energy/radiance/repetition-rate. Fig. 9 is a schematic representation of this system showing the two lasers mounted on a Bridgeport Series II Milling Machine. A helium-neon laser is incorporated for aiming purposes, as well as providing light for a focus device. The focus device measures the distance from the focus lens to the work piece and maintains the required focal relationship within ±0.004". This machine is a 5-axis device capable of drilling holes in all but the very largest components of jet engines and has been used on a three shift per day basis drilling these parts. Fig. 10 shows a cross section of a hole 0.020" diameter (exit) through 0.120" thickness cobalt base alloy. The complete absence of recast material which should be noted in this section is due to the previously mentioned characteristics of glass lasers and also special techniques developed for this particular application. The hole can be seen to have smooth side-walls and it's felt that this may result in a longer lifed part as it is less likely to develop thermal-stress cracks. The hole shown is produced in 7 seconds at an operating cost of less than 1¢. Holes of this diameter have been drilled through material thicknesses of greater than 0.65" on a production basis. An overall photo of one of these systems is shown in Fig. 11 with the drilling station contained behind an interlocked safety enclosure. In addition to the two laser power supplies, an additional rack for the computer system is seen. The computer interfaces with all components of the system and is used for both automatic and manual operation. It is capable of reading punched-paper-tape, as well as accepting manual commands directly from a keyboard. When operated from the paper tape input, each set of parameters for each hole is displayed on the CRT. These parameters are:

X, Y, Z, ϕ, & θ position
Z correction (+ or - for best hole drilling)
Energy setting of the laser power supply
Number of laser shots for the hole

Any step of the sequence can be changed by stopping operation, keying-in desired changes, and then allowing the edited program to run. New tapes, of the edited version, can be supplied from the computer upon command.

The computer is capable of storing sufficient information to drill up to 500 holes (all with 5-axis motions required) in this edit mode of operation.

New systems similar to the one described are being quoted at approximately $150,000.

THE FUTURE

Presented here have been some examples of current production laser drilling machines and the characteristics of their glass lasers. The contrasts of YAG with glass laser materials and, indeed, all mention of the glass rods here alluded to conventional silicate-base glasses whose characteristics have not changed for at least the last five years. As was seen, this material has significant advantages for use in laser drilling. There are, however, new glasses in development and some of these are even better. It now appears that phosphate-base glass will, within the next year or so, be coming into great use. The phosphate glasses have much better characteristics than the silicates, as can be seen from Table 1. It is expected that the use of new materials, and further advances in laser drilling techniques will result in many more industrial uses of this method of making holes.

TABLE 1

Phosphate - Glass Characteristics Relative to Silicates

Characteristic	Performance
Efficiency	1.5X
Radiance	1.1X
"Spiking"	≃Same
Repetition Rate	≃2X

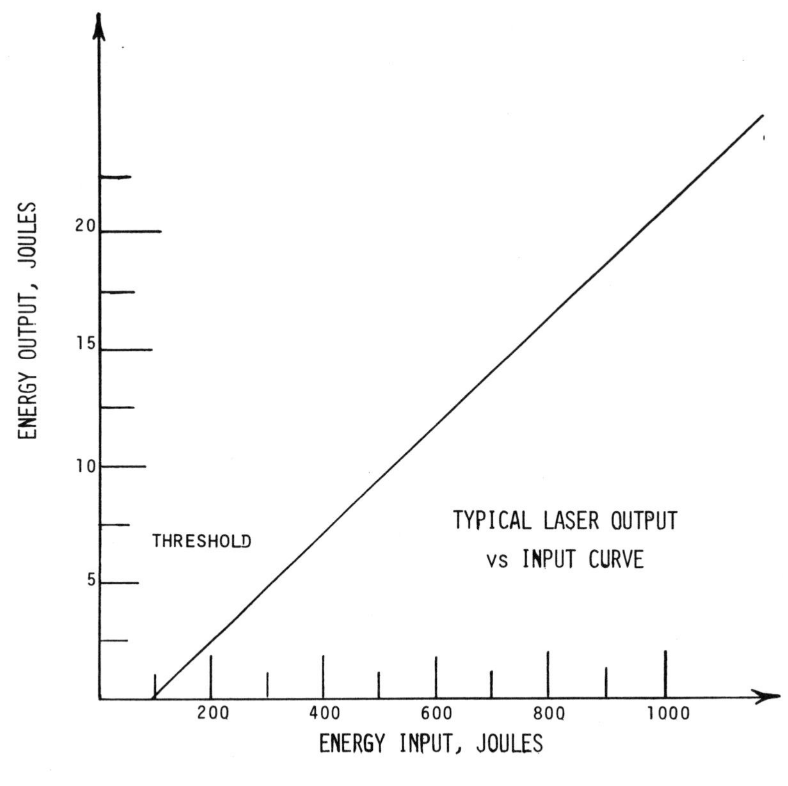

FIGURE 1

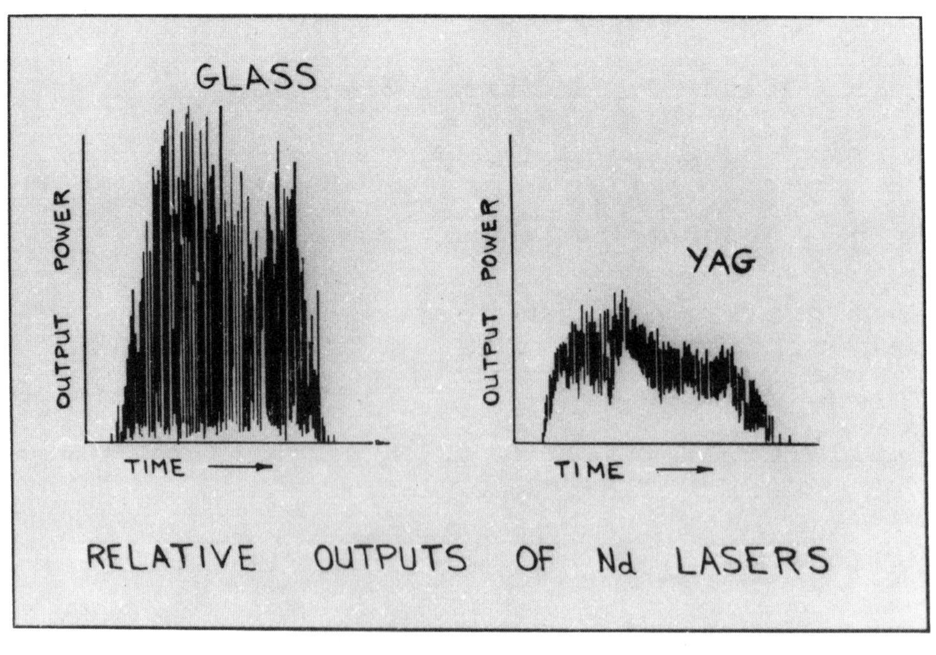

FIGURE 2

FAR FIELD IMAGING

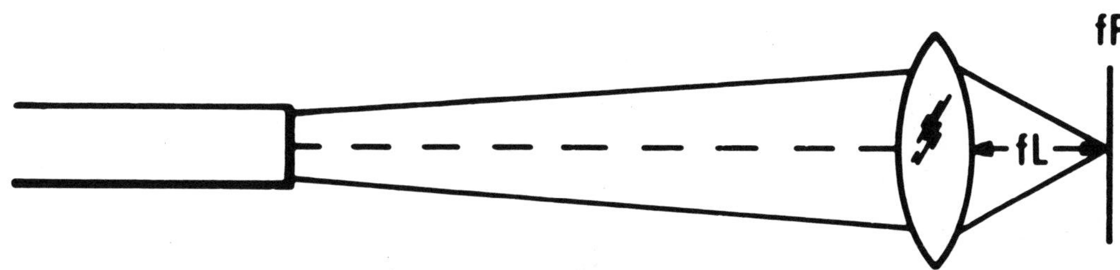

SPOT DIAMETER = θfL

θ = BEAM ANGLE [RADIANS]

FIGURE 3

NEAR FIELD IMAGING

DIAPHRAGM OR ROD END

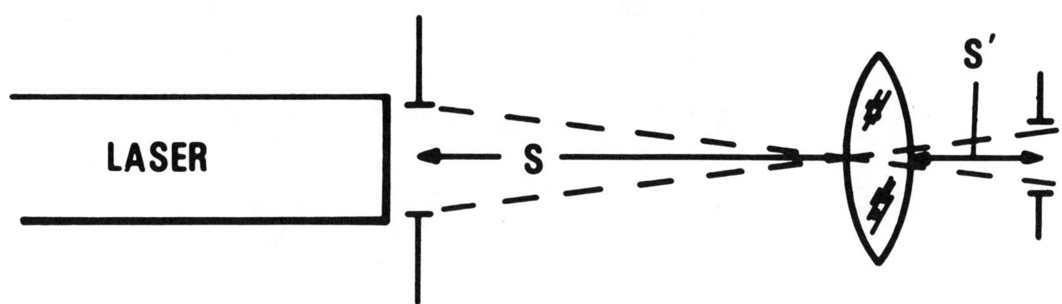

1/S + 1/S' = 1/fL

SPOT SIZE = DIAPHRAGM DIAMETER X S'/S

FIGURE 4

FIGURE 5

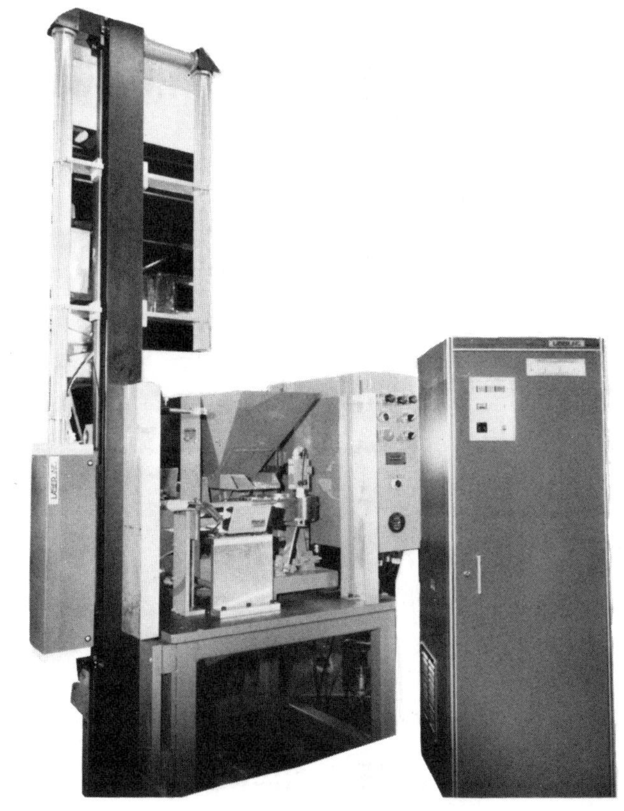

FIGURE 6

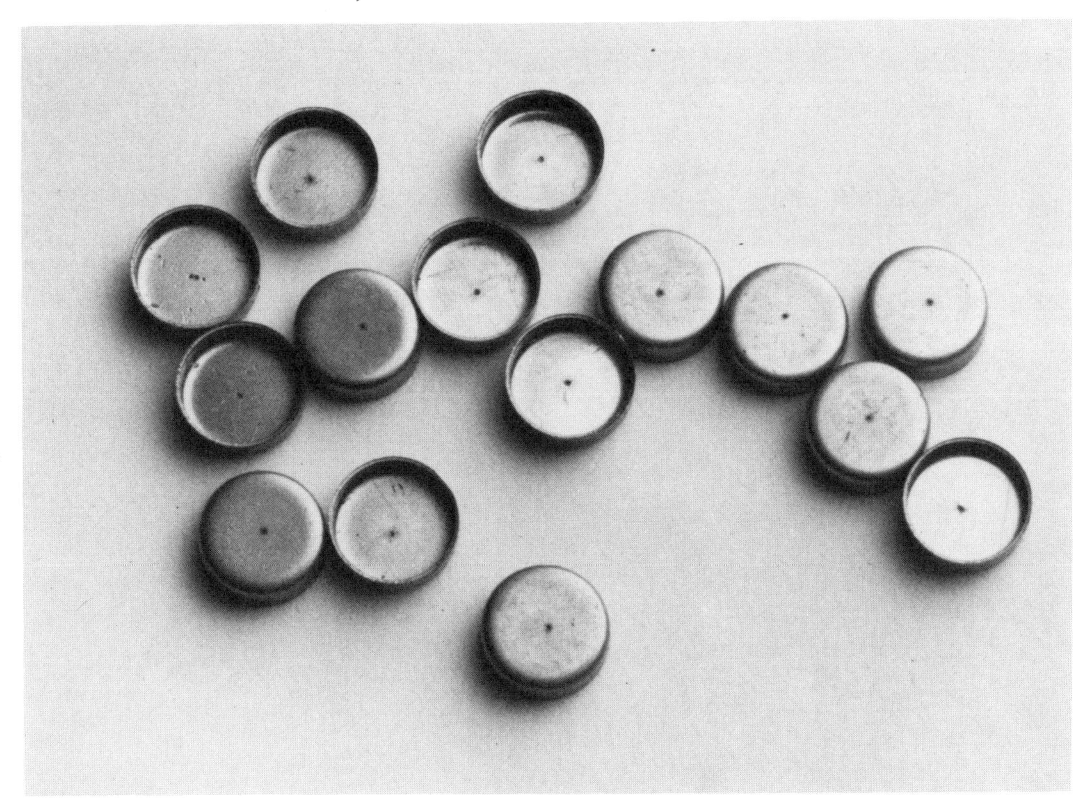

FIGURE 7

FIGURE 8

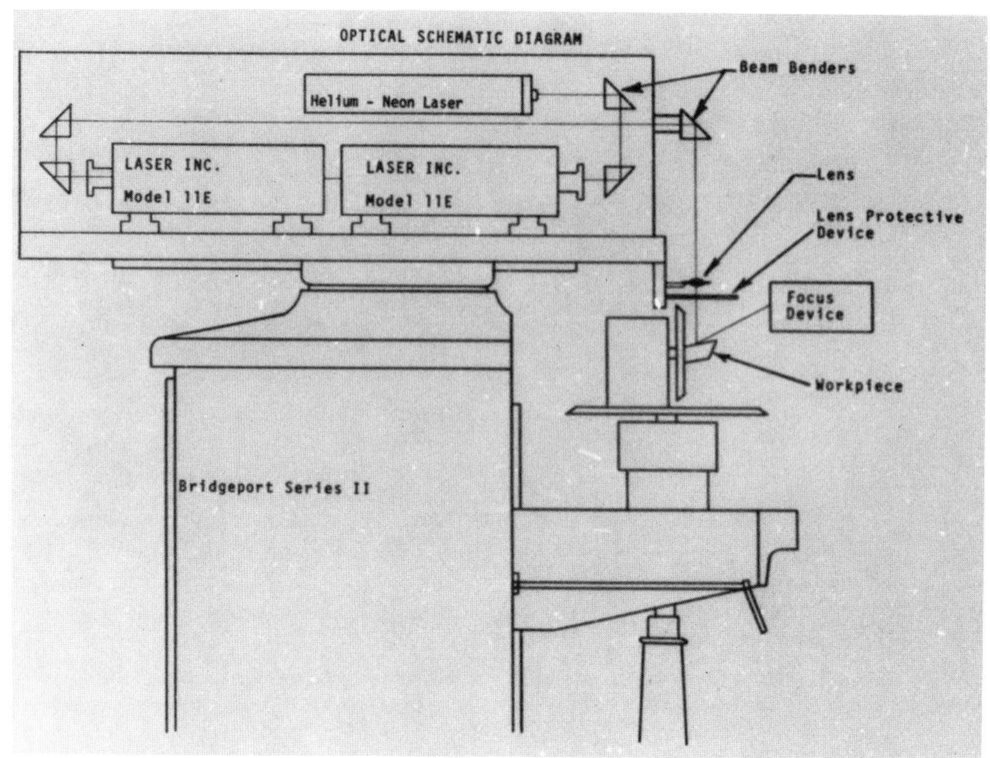

FIGURE 9

FIGURE 10

FIGURE 11

Precision Metal Cutting With Carbon Dioxide Lasers

By David Belforte
Ferranti Electric Inc.

Medium power Carbon Dioxide lasers provide a unique combination of low operating cost and improved results in the process of profile cutting sheet metal. The lasers ability to produce a sharp, clean edge at a high cutting rate with a minimum set-up time affords users an opportunity to reduce the costs associated with limited volume sheet metal fabricating operations. Illustrations of several successful carbon dioxide laser installations will show the cost effectiveness of laser cutting over conventional techniques.

Introduction

One of the most basic operations is a manufacturing plant is the sheet metal fabricating shop. Here sheet metal is sheared, slit, punched, notched, nibbled and sawed to produce a variety of parts for further assembly. Traditionally a labor intensive operation, the sheet metal shop became the subject of intensive cost reduction studies several years ago. High speed slitters, shears, punch presses and nibblers increased the shop output and semi-automatic and numerically controlled machines reduced the labor intensive operations further. However one segment of the fabricating shops operations continues as a high cost operation - custom or limited volume production of complex shaped parts. It is this manufacturing problem which carbon dioxide lasers have begun to solve. This paper will outline the advantages of laser metal cutting which result in its current acceptance as a cost effective alternative to conventional sheet metal cutting operations.

The Carbon Dioxide Laser

Among the many lasers used in commercial applications today one, the carbon dioxide (CO_2) laser, offers the metal processor maximum benefits. Lasers are electro-optical devices which produce a highly collimated, small spot size, single wavelength beam of light of extremely high energy. This light energy when caused to impinge on a surface is absorbed by the target material (to some degree) and the energy is converted to power in the form of work.

The theory of lasers has been amply described in the technical and laymans literature. For the purposes of this paper it will suffice to say that certain materials when caused to "lase" produce a beam of light at a specific wavelength. The most commonly used laser, helium/neon, produces a visible red beam of light whose wavelength is 0.6328μm (micro-meters) at extremely low power outputs in the range up to 1.5 milliwatts. The laser discussed in this paper, the CO_2 laser, produces an invisible beam of light whose wavelength is 10.6μm. CO_2 lasers have been developed which can generate many kilowatts of power. Of interest to the fabricators of sheet metal are the medium power CO_2 lasers ranging from 350-1000 watts.

The output power of a medium power CO_2 laser is directly proportional to the length of its resonator cavity. Figure 1 illustrates a typical medium power CO_2 laser. The resonator cavity being the gas container (resonator tube) and the end mirrors. The output power is produced through the generation of photons developed by the excitation of CO_2 molecules in the resonator by a high voltage discharge. A conventional flowing gas CO_2 laser will develop 50 watts of output power per meter of resonator cavity length. Thus a 500 watt laser requires a resonator cavity 10 meters in length. This would result in a resonator structure of unwieldly length so laser manufacturers, using mirrors, developed the folded resonator concept. By use of accurately positioned mirrors the optical path of the resonator is of the required length while the overall length of the structure is minimized. A unique version of a folded 500 watt CO_2 laser is shown in Figure 2. In this version the resonator tube is folded many times through the use of thermally stabilized mirrors to produce a lightweight, compact structure. Other manufacturers employ thermally stable base plates such as granite slabs to achieve the same purpose - rigid, optically aligned mirrors.

Often noted in CO_2 laser specifications is the term TEM_{00}. TEM (Transverse Exitation Mode) refers to the laser oscillation phenomenon. An optically stable laser cavity oscillate along the axis of the cavity thus the defining term is TEM_{00}. As larger amounts of energy are pumped into the laser gas some lasers will oscillate in other modes, such as TEM_{01}. Although satisfactory cutting performance has been attained using the higher order mode structures the optimum results are obtained using the TEM_{00} mode. TEM_{00} lasers produce a beam with a smaller divergence than lasers with higher order modes. The greater flexibility of the TEM_{00} mode structure is shown by the fact that the smallest spot size for a given depth of focus can be achieved. Conversely the greatest depth of focus for a given focussed spot size can be achieved. Therefore the absolute minimum spot size that can be achieved with simple lenses, before lens aberrations begin to dominate, is also smaller. A TEM_{00} mode laser in a metal cutting application will produce narrow, straight sided cuts with a minimum heat affected zone and negligible slag.

The CO_2 laser requires a precise mixture of three gases, carbon dioxide, helium and nitrogen for high power output operation. The basic gas and the one which supplies the molecular action required for photon generation is carbon dioxide (CO_2). Nitrogen (N_2), an inert gas, acts to sustain and reinforce the molecular action. Because the heat generated by the gases in the lasing action are of such a high temperature that the resonator tubes could melt, helium (He) is added to the mixture as a cooling agent. The mixture of these gases in a typical medium power CO_2 laser is; CO_2 - 5%, N_2 - 15% and He - 80%. In a flowing gas system this mixture is continually pumped through the resonator to sustain the lasing action, thus gas consumption becomes a factor in multiple shift operating cost. One manufacturer has developed a patented system to regenerate the gases, reducing his consumption to an acceptable level.

The Laser Beam Delivery System

Once the laser action develops to a certain level, portions of the photon energy are released through a partially reflecting cavity mirror (Figure 1) as an unfocussed or raw beam of a diameter dependent on the cavity design. This raw beam, although containing the full power output of the CO_2 laser, has a diameter large enough so that the energy per unit area is not sufficient to produce the instantaneous high power required for metal cutting. Therefore the raw beam is brought to a focus by a condensing lens which is selected to produce a desired spot size and depth of focus.

A typical optical delivery system for a medium power CO_2 laser is shown in Figure 3. Optical delivery systems can be attached directly to the laser or they can be located remotely as in the case of moving optics. In the operation of very high power CO_2 lasers, not discussed in this paper, reflecting optics are used because refracting materials cannot withstand the high power generated.

The focus lens materials most commonly used in metal cutting applications are gallium arsenide and zinc selenide. The former requires water cooling of the lens holder while the latter can be air cooled, simplifying the lens holder design. Either material will deliver long life if simple preventive maintenance procedures are followed. Experience has shown that focus lens replacement cost, they are relatively expensive items, can be minimized by periodic examination and cleaning of the lens if required. Many industrial users maintain spare preloaded and aligned lens holders so that a quick changeover can be accomplished for routine maintenance.

Gas Jet Assist

As stated earlier the laser beam when caused to impinge on a surface will be absorbed to some degree. A determination of the ability of a material to absorb the 10.6μm wavelength of the CO_2 laser can be obtained from electrical conductivity tables because absorbtion is inversely proportional to electrical conductivity. Thus high conductivity materials such as gold, copper and aluminum are poor absorbers while plastics and wood are almost perfect absorbers. Absorbtion is not the only determining factor in laser cutting. Some materials have surfaces which will reflect the CO_2 laser light at room temperature. Among these are the materials of interest to the sheet metal fabricator. A series of tests using a 1 KW CO_2 laser showed that the threshold for metal cutting is reflectivity dependent. The extent of the reflectivity depends on the surface finish. Typical reflectivities are; stainless steel 77%, aluminum 95% and nickel alloy 83%.

It was found that by introducing oxygen at the point of laser beam impingement, cutting of steel could be enhanced. The oxygen jet serves several purposes. First, it acts to reduce the surface reflectivity so that more efficient laser beam absorbtion takes place. Once the steel surface reaches a high temperature the oxygen acts to initiate an exothermic reaction which causes a molten puddle to form which then grows until a hole is pierced in the piece and the cutting process begins. The gas flow which is confined to an area only slightly larger than the laser beam diameter acts to cool the rest of the metal giving rise to a very narrow cut. Further the gas flow directs vaporized metal from the cutting area insuring full impingement of the laser beam on the steel. At the bottom of the cut the flow of gas sweeps away molten slag producing a relatively dross (slag) free cut.

The process of gas jet assist laser cutting bears a strong resemblance to flame cutting processes such as oxyacetylene cutting. The essential difference is that in oxyacetylene cutting the flame diameter is typically 0.125". Therefore the cut width will be of this size. In laser cutting with gas jet assist confined to an area slightly larger than the beam the cut width is typically 0.008".

When optimum laser cutting conditions are achieved the balance of heat supplied by the laser and the exothermic reaction are such that only the metal traced by the focussed laser beam spot is cut. In this situation approximately 30% of the heat is supplied by the laser and 70% by the reaction with oxygen.

It was through the development of gas jet assist cutting that the medium power CO_2 laser became an effective sheet metal cutting technique. A typical industrial laser metal cutting operation is shown in Figure 4 where the laser beam is directed to the sheet metal through a gas jet assist nozzle device.

Use of the gas jet assist cutting requires the optimization of the gas nozzle design. It has been shown that the pressure of the gas flowing from the nozzle will have a direct bearing on the quality of the cut. A low pressure flow produces a self-burning effect which can be corrected by raising the pressure to the point where the laser spot diameter controls the width of cut. However at this point the parent metal is undercut and a thick recast layer (heat affected zone) and heavy dross are present as shown in Figure 5A. Better results are obtained using a gas nozzle which produces a converging jet at sonic velocity. This design improves the cut quality appreciably by producing a thin uniform heat affected zone and a minimum dross as shown in Figure 5B.

Of equal importance to the sheet metal cutting process is establishment of a proper gap between the gas nozzle and the workpiece. An analysis of gas flow conditions for various pressures and gap spacing have been conducted so that manufacturers of laser cutting systems can now recommend settings consistant with good results for a range of metal thicknesses. In practice, when a short focal length lens is used, a gap of between .035" and .050" is desirable. Maintenance of this gap spacing is important so manufacturers provide devices such as the self-adjusting height sensing unit shown in Figure 4 which controls the gap spacing automatically regardless of the surface unevenness commonly found in sheets of steel.

In addition choice of lens focal length and determination of the placement of the focussed spot are variables which are considered. As an example low carbon steel, 0.128" thick is best cut using an f2.5 lens with the spot focussed at the metal surface. Stainless steel of the same thickness shows the best results when an f2.5 lens is focussed just below the surface.

Sheet Metal Cutting Systems

The first consideration to be made in choice of a sheet metal cutting system is the size of the material to be cut. In a conventional fabricating shop the most common requirement is to cut piece parts from stock sizes of sheet which range from 4' x 8' to 6' x 15'. Some shops work with rolls of sheet metal which is supplied in widths up to 4'. Large manufacturing operations may work with pre-sheared sheet stock of a more manageable size typically 3' x 3'.

In any case a decision on the type of system motion is required. Basically the user has a choice of three approaches, beam motion, work motion or a combination of these. Beam motion can entail movement of the entire laser over the workpiece, movement of the optical delivery system or movement of the laser beam only. Work motion requires that the work move in the X and Y directions under a fixed beam, usually by means of a movable work table. A combination of work table motion in one direction coupled to laser or optic motion in the other is a very useable approach to certain processing problems.

Manufacturers of laser cutting systems employ any of the above motions which will produce the most effective and cost justifiable solution to a users cutting operation. The most common use is movement of the laser optics over a stationary workpiece, as this procedure reduces the amount of floor space required to move a piece the full distance in the X and Y direction.

The type of motion system chosen for the application depends on the range of cutting speeds anticipated and the accuracy of the parts to be produced. A variety of motion system components can be used including; chain drive, rack and pinion, ball screw or friction. Each can provide smooth, constant speed motion with varying degrees of accuracy. It must be remembered that once the laser beam is properly aligned to the work it will not change its pointing location. Thus the accuracy of the cut is wholly dependent on the accuracy of the drive system. Here again the type of metal processing requirement will determine which drive mechanism applies. For example air conditioning duct work piece parts need only be cut within a ± 0.030" tolerance for many applications, thus the motion system need only meet this accuracy requirement. Stainless steel counter tops however are normally fabricated to a ± 0.015" tolerance so a chain or friction drive system would not suffice. Precision metal parts which require tolerances of ± 0.001" would naturally require a very precise motion system such as ball screw or rack and pinion.

The control of the motion system is an important factor from both an accuracy and cost standpoint. Control speed is not a major factor in sheet metal processing as cutting rates are typically under 250 inches/minute for most operations. This does not preclude a certain number of metal cutting applications which when processed on high speed tables with a 1 KW laser result in very high speed cutting. However the subject of this paper, sheet metal profile cutting with medium power CO_2 lasers, is confined to moderately priced systems which incorporate lower cost motion systems.

Motion control is usually accomplished by use of an optical follower tracing device or by a numerical control input, either tape or computer. The lowest initial cost results from the optical follower approach. Here a sensing head follows the pattern represented by a line drawing on a 1:1 scale. The optical device operates by sensing the difference between a black line and the white paper. The two most common units are an edge follower and a line follower. Each has its own advantages for a given trace profile and preference depends on the users inclination. Optical follower tracing systems are mostly used with conventional gas cutting machines where the laser or optical delivery system replaces the cutting torch. One version of a cantilevered cutting system is shown in Figure 6. Here the part drawing is placed on a template table, the optical head is positioned over the drawing and once initiated proceeds to follow the drawn profile. The output from the tracing head is electronically amplified into signals driving the X and Y motors which position the cantilevered arm and the laser causing the laser beam to follow the pattern, producing a duplicate cut in the workpiece. This type of system will have an accuracy of ± .015" over a 4' x 4' table.

Numerical control of the motion system produces a more accurate cut pattern of very complex profiles. Tapes are prepared by either writing programs for the cut pattern or by using a digitizing system to generate the coordinates. In highly repetitive operations or in instances where the same requirement occurs over a period of time, tape storage is a positive cost factor. A tape controlled laser cutting system is shown in Figure 7. Here the tape controlled drive motors position the work in one direction and the laser in the other. The system has an accuracy of $\pm$ 0.001 over a table size of 4' x 4' at cutting rates up to 250 inches/ minute.

Computer control of a laser cutting system is currently not a viable technique due to costs associated with the operation. However consideration is being given to this approach for several promising applications and it is conceivable that this form of control will become more popular in the future.

Auxiliary equipment required for a shop installation of a laser cutting system are; a source of cutting assist gas, a cutting table to support the work, an exhaust system to vent noxious fumes and approved safety devices. The oxygen or air used as a cutting assist gas is fed into the gas jet assist nozzle from a regulated supply. The procedure used in a typical cutting system is to actuate the gas flow a few seconds prior to laser beam initiation so that the cutting process occurs under protective gas flow conditions. Cutting tables vary from one user requirement to another. The simplest arrangement shown in Figure 6 uses knife edged aluminum bars which support the sheet metal while allowing slag and fumes to pass through. The knife edge of the bars withstand the effect of the laser beam for long periods of time and are easily replaceable if damaged. Other support systems such as movable copper pins or honeycomb panels have been used successfully. Exhaust systems are required to remove noxious fumes from the cutting area. Some exhaust systems are part of the work table, as shown in Figure 7. In this arrangement the sheet metal on the work table surface acts as a cover so that the underside of the worktable works like a plenum chamber insuring good exhaust flow conditions.

Operator safety is a major consideration in any industrial machining installation. The CO_2 laser cutting system requires operator protection in three areas; high voltage, radiation and physical damage. Federal and state agencies have been charged with developing regulations covering all of these areas. Under the OSHA regulations the user is required to perform specific safety procedures. Laser manufacturers are very familiar with the safety requirements and they can provide the user with information and recommend safety precautions. When these precautions are taken the laser becomes a safe tool in the workshop.

Benefits of Laser Metal Cutting

The medium power CO_2 when used in a sheet metal cutting application offers a number of important advantages.

- Material waste (kerf) is held to a minimum because the cut width is in the range of 0.006" - 0.015"
- Minimum preparation (marking-out, degreasing, clamping, etc.) is required
- There is no wear on, or replacement of cutting tools - a particular advantage with hard to cut materials
- A smooth edged cut can be obtained at high cutting speeds

- Total heat input is low, therefore, there is very little distortion or damage to heat sensitive materials
- A parallel sided cut is made
- There is in "tool-material" contact, therefore, no mechanical distortion or surface marking when cutting thin materials
- Material upset (dross) is minimal, reducing or eliminating clean-up operations
- Sharp contoured profiles can be cut - the laser beam is not sensitive to changes in cutting direction
- Hardened steel can be cut with no resulting annealing of the parent metal
- The cut can be made without a starting or stoping tab therefore cuts can be started away from the edge of the piece

All of the above advantages represent a benefit to the sheet metal fabricator for each of them provides for the elimination of a cost usually associated with conventional metal cutting. The items listed are self explanatory and users considering CO_2 laser cutting can factor the resultant savings into their cost of operation.

Because of the potential for high speed cutting many users consider cutting rate as a major factor in cost justification of the laser. Although rate is a factor experience has shown that the majority of successful, cost justified laser cutting installations found other factors to be more important. The ability to cut a sheet metal part at 100 inches/minute rather than at 25 inches/minute is certainly a cost benefit. More important however are the savings experienced by elimination of the costly secondary operations after cutting. These operations, usually manual, such as edge grinding, straightening, surface refinishing or annealing are time consuming and one of the major cost factors in the production of limited volume piece parts.

In many sheet metal shops the appearance of the cut edge produced in mild steel by the CO_2 laser is of such a quality as to be used as is. The fine striations exhibited are normally acceptable, precluding any finish grinding operation. Laser cuts in thin gauge stainless steel result in a heat affected zone of .008" thickness. A light grind removes this layer, if necessary, however if the next operation is welding, this layer is usually taken up in the weld puddle and is not detremental to the weld strength. Laser cutting of galvanized steel does not result in zinc depletion so that the surface of the piece retains its protection up to the edge of the cut.

A comparison of the cost of laser cutting varies depending on the CO_2 laser used. One laser manufacturer reports a cost of $0.04/linear foot of cut - a figure which includes all consumable costs as well as labor and overhead, but not amortization. Other laser manufacturers costs may be somewhat higher but still represent a significant savings over conventional mechanical cutting costs.

In a recent cost justification study a metal fabricator indicated that the greatest savings he anticipated were in the elimination of tooling costs he normally experiences with dies. Although the laser rate was a factor, the savings in tool manufacturing, maintenance and storage costs far outweighed the time saved in cutting.

Cutting Applications

The list of successful CO_2 laser metal cutting applications is growing as the aforementioned cost benefits are made known to potential users. Among the more significant applications are the following which were chosen to illustrate a specific benefit which results in the choice of laser cutting over conventional methods.

- A manufacturer of custom electrical enclosures uses the CO_2 laser to produce cutouts in painted panels for meters, gauges, louvers, etc. Formerly, hole punches and friction saws were used requiring surface masking, manual layout and post-cutting clean-up. One typical job showed a reduction of eight man-hours using the laser.
- A supporting structure for a rotor blade assembly is produced by laser slitting mill supplied stainless steel to the proper width and then using the laser to trim the subsequently stretch formed part. A cost savings of 80% over conventional slitting, milling and grinding results, with the major portion of the savings gained through elimination of tooling costs.
- Duct work piece parts are laser cut from a continuous roll of galvanized sheet. The optical follower controlled laser cutting system produces parts to a finished dimension at a savings of over $45,000/year. The laser system replaced lengthy marking out procedures, manual shearing and costly edge clean-up and straightening.
- Stainless steel petrochemical seal rings are laser cut from sheet stock so that maximum utilization of the sheet occurs. The lasers ability to cut shapes in close proximity and its ability to start anywhere on the trace allows nesting of parts so that scrap is reduced. The operating cost of the CO_2 laser on this job is low enough to provide a piece part selling price 1/3 that of conventionally rotary sheared parts.
- Fuel tanks of custom design for military vehicles are produced in small quantities and are fabricated from complex shapes. The cost of custom cutting the steel in a limited volume requirement is dramatically reduced through use of a laser cutting system which is controlled by the optical tracing of assembly drawings. Further the cut edges are of a high quality, eliminating the need for grinding prior to subsequent welding operations.
- Stainless steel signs featuring highly stylized letter cutouts are produced by laser cutting the profile traced by an optical follower positioning system. Laser cutting results in one step process, eliminating layout, masking, friction sawing and edge clean-up. The fabricator is also able to reproduce sharply profiled cuts which are impossible to produce using mechanical cutting techniques.
- The production of small diameter holes in thin wall steel tubing is facilitated by use of the laser beam in a trepanning operation. In this operation the laser beam traces the hole diameter so that the laser

beam cuts through without sensitivity to angle of approach. Because laser cutting is a non-contact technique the need for supporting mandrels and fixturing is eliminated. Further mechanical operations result in edge burring leading to post cutting clean-up operations.

Summary

This paper has been presented to show the developing acceptance of the laser in the workshop. The unique characteristics of CO_2 laser metal cutting produces a number of cost benefits to the user resulting in improved productivity and cost reduction. Through the use of gas jet assist cutting the laser has become an effective alternative to conventional sheet metal processing techniques.

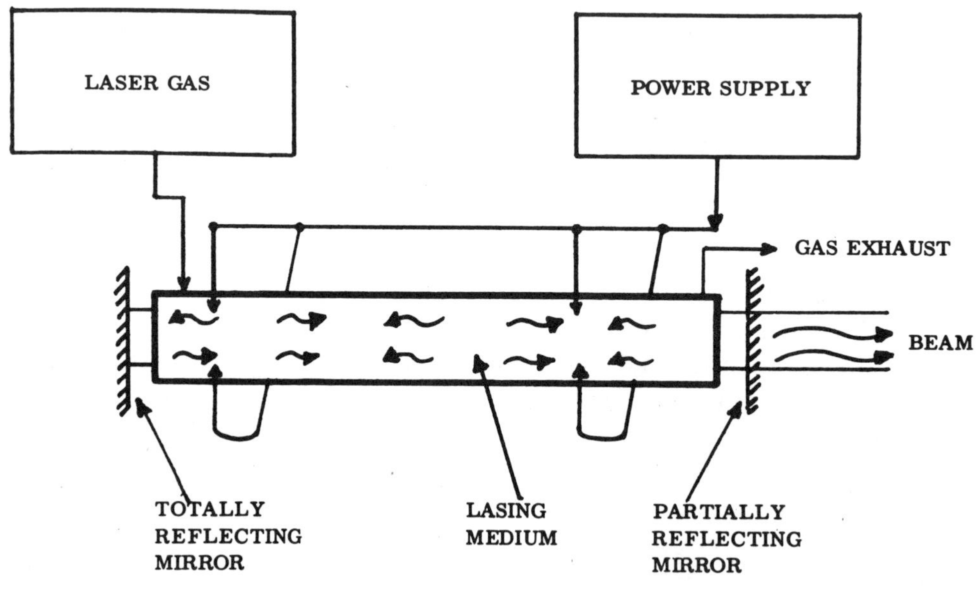

Figure 1 Diagram of CO_2 Laser

Figure 2 Compact, multi-fold, 500 watt CO_2 laser

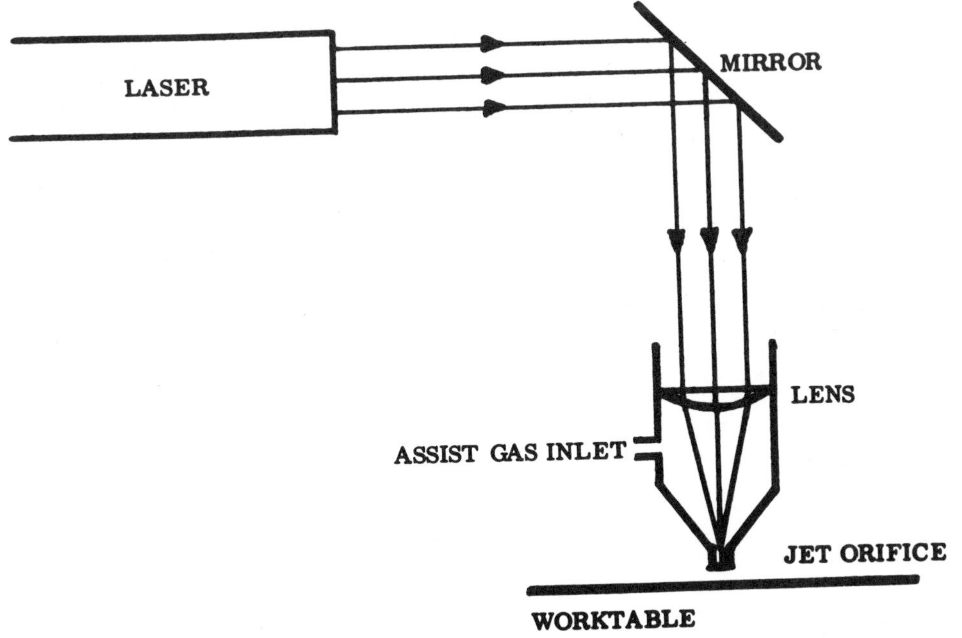

Figure 3 Typical laser beam optical delivery system

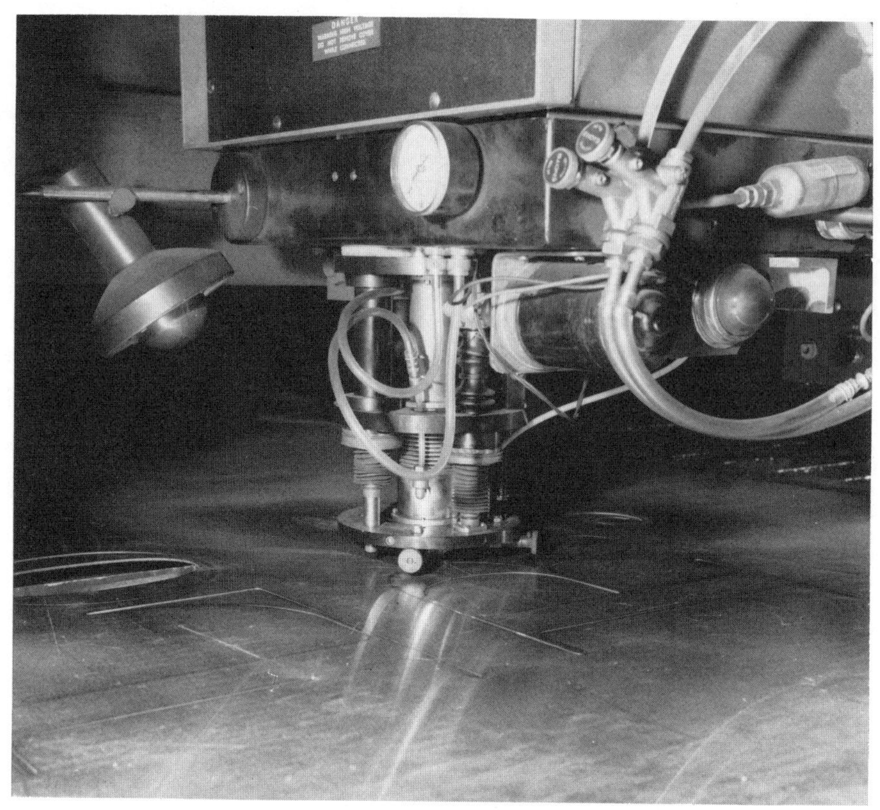

Figure 4 Optical delivery system with automatic height sensing device

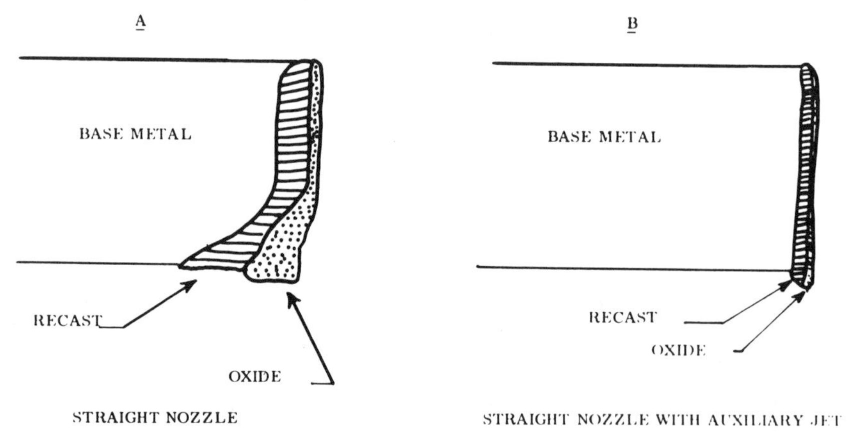

AFFECT OF GAS JET ASSIST ON
EDGE CONDITION

Ferranti Electric, Inc.
Sturbridge, Mass.

Figure 5 Effect of gas jet nozzle design on cut quality

Figure 6 500 watt CO_2 laser adapted to conventional optical follower positioned gas cutting machine

Figure 7 Split axis laser cutting system operated by numerical control

Laser Cutting Of Thin Materials

By Simon L. Engel
Electronic Systems Group, GTE Sylvania Inc.

This paper summarizes some experience in Laser Assisted cutting of thin materials (0.375 inch or less). It has been found that cutting rate is inversely proportional to the thickness of the material, while directly proportional to the input power level beyond a certain threshold value. Use of Reactive Gases increases cutting speeds in chemical reaction dominated processes. Self-burning of the material, however, presents kerf quality problems and eventually defines the lower limits to the cutting rates. Use of Non-Reactive Gases produces the best quality cuts. Data is presented on both metallic and non-metallic materials.

INTRODUCTION

Thin materials, 0.375 inch or less in thickness, fill an important role in present day technology. Structural members formed from thin materials have a high strength to weight ratio as well as low manufacturing cost. Since most materials are supplied in a sheet or strip form, the manufacture of the final product involves several steps of cutting, shaping, welding and trimming. Material working with lasers has been considered ever since the introduction of lasers in the early 1960's. Lasers were to provide the high concentration of energy needed in material cutting and welding on a mass production basis. Several years went by, however, before sufficiently powerful, reliable and low cost lasers were developed to enter the commercially oriented manufacturing field. Today, lasers are available up to 16 kwatt output, in the price range of $10,000 to $500,000. This paper addresses itself to laser assisted cutting and the manufacturing capacity of a variable power, 1 kwatt maximum, TEM_{oo} Mode CO_2 laser. Lasers in this power range cost approximately $60,000.

TEM_{oo} MODE LASERS FOR CUTTING

In most non-laser assisted electrothermal cutting processes the material is brought to a melting point and is then blown away by a jet of gas. In laser assisted cutting the aim is to vaporize the material very quickly by concentrating the laser energy into a very small spot. If the vaporization is achieved rapidly enough, little heat will be conducted into the material, resulting in a narrow heat affected zone (HAZ) and no physical distortion of the work material. TEM (Transverse Excitation Mode) refers to the ability of a laser to oscillate in the laser cavity along one or more paths parallel to the axis of the cavity. The subscripts $_{oo}$ indicate the fundamental mode, i.e., lasing occurs only along the axis of the cavity. It is not easy to assure TEM_{oo} type operation of a high power laser, since the laser has a tendency to oscillate in some other mode such as TEM_{o1}, TEM_{o2}, as greater amounts of energy are pumped into the laser medium. Even though the total power of a "multimode" beam may equal that of a TEM_{oo} beam, the power density at its focal point may be up to two orders of magnitude less than that achievable with a TEM_{oo} beam. TEM_{oo} laser beams can be focused to the smallest spot theoretically possible, typically a few mills in diameter. A 1000 watt TEM_{oo} laser beam focused through a 2.5 inch focal length lens may achieve 10^{+8} watts/in^2 at the focal point, enough to vaporize even Tungsten which requires 2 x 10^{+6} watt.second/in^3. From the manufacturing point of view then, the use of a TEM_{oo} mode laser means narrow kerf widths, square kerf sides, narrow or no HAZ, minimum recast layer and minimum underside slag. In actual practice kerf widths of 0.005-0.015 inch have been consistently achieved with the focal point positioned about 1/3 of the material thickness below the surface using an 0.050 inch diameter gas jet nozzle, positioned about 0.015-0.020 inch above the surface. Selection of the gas depends on the material to be cut, and is noted along with the data.

CUTTING OF THIN METALS

All metals are relatively good (80%+) reflectors of 10.6 um wave length (infrared) energy at room temperature. It would seem then, that in the case of exceptionally good reflectors such as aluminum and copper, there may not be sufficient laser energy absorbed by the material to initiate vaporization or even melting. Several investigators, (1,2) however, have predicted by theory, or reported by experiment, a substantial increase in absorptivity of metals at elevated temperatures. The reduced thermal conductivity of metals in the molten state is considered to be one important factor. By theory, molten Aluminum, Copper and Silver approach the absorptivity of Stainless Steel at room temperature. While no data is yet available on the absorptivity of the two phase mixture of molten metal and its vapor, our experience and some approximate calculations indicate that absorption may be as high as 50 per cent. This increase of energy absorption over theoretical makes possible the cutting of Aluminum as documented elsewhere in this report. While there must be sufficient energy density to produce the vapor state, in practice this also means precise focusing. Focusing becomes progressively more critical for a given material as its thickness is increased. It is also critical for very high melting point materials. In the case of cutting 0.020 inch thick Tungsten, for instance, focusing accuracy must be within ± 0.003 inch

---

1. T. J. Wieting and J. T. Schriempf: Free-Electron Theory and Laser Interactions with Metals, Report of NRL Progress, June 1972.
2. J. E. Harry and F. W. Lunau: Electrothermal Cutting Processes Using a CO_2 Laser. IEEE Transactions on Industry Applications, Vol. 1A-8, No. 4, July/August 1972.

(2.5 inch focal length lens). In the case of Titanium of the same thickness, it is ±0.020 inch.

Further verification of high absorption is concluded from our experience that surface finish and surface oxides have little or no apparent influence on the cutting rates.

EMPIRICAL CUTTING PARAMETERS

The observed cutting rates for both metals and non-metals have been found to fit the following empirical relationship:

$$\text{Cutting Rate } \left(\frac{\text{inch}}{\text{min}}\right) = K \frac{P}{E \cdot A \cdot t}$$

where P = laser power incident on surface (watts)
E = ablation (vaporization) energy of the material $\left(\frac{\text{watts}}{\text{in}^3}\right)$
A = area of laser beam at focal point (inch^2)
t = thickness of material (inch)
K = a constant, characteristic of the material and the coupling efficiency of laser energy into the material $\left(\frac{\text{inch}}{\text{min.}}\right)$

Thus, cutting rate is inversely proportional to the thickness of the material and is directly proportional to the input laser power above a certain threshold. This latter fact means that the cutting rate achieved by one 1000 watt laser beam cannot be equalled by two 500 watt or four 250 watt beams. (The 250 watt focused beam may not even reach the needed threshold). This concept of threshold must be considered when beam splitting and/or several simultaneous work stations are contemplated to be designed into a system.

THE CUTTING OF THIN METALS

A. Chemical reaction assisted cutting.

Cutting rates of Carbon Steel and Titanium in Figures 1 through 4 are examples of the cutting rates possible when the heat due to the exothermic reaction of these metals with oxygen adds to the absorbed laser energy. These cutting rates are typically 40% higher than is possible when inert gases are used.

For maximum productivity, lasers are often part of automatic tooling or NC/DNC positioners. Although the cutting rates plotted are maximum rates, there is a range of slower rates possible for any one given power level, where acceptable quality cuts are produced. Thus, if an X-Y positioner should slow down while making a sharp turn, it may not be necessary to reduce the laser power level. However, in the case of chemical reaction assisted cutting there is a lower limit of cutting rate at and below which "self-burning" occurs. The material burns without the aid of the laser and the kerf becomes wider, erratic and rough. Figures 5 and 6 show the Upper and Lower Limits of cutting rates for Carbon Steel and Titanium at 1000 watt power level. The self-burning may be avoided and good edge-condition assured, of course, by the use of an inert gas at the expense of reduced cutting rates.

B. Non-reactive metal cutting.

Laser assisted cutting of Aluminum and (the Nickel-based super-alloy) Hastalloy-X are examples of this class.- Figures 7 and 8. Aluminum is of high economic interest to the aircraft industry where N.C. controlled laser cutting of intricate aeroplane parts is being investigated to replace expensive blanking dies.

THE CUTTING OF THIN NON METALS

Laser assisted cutting of quartz, Figure 9, is an excellent example of manufacturing cost reduction potential. The cutting rate of 5 inches/minute for 0.375 inch thick quartz is approximately 100 times faster than that obtainable with saws. Several manufacturing steps may be omitted as well, by cutting the final geometries of lenses and windows directly. As an example, lens blanks cut from 0.096 inch thick stock showed a kerf width of 0.015 inch and a crack-free 0.030 inch HAZ, as identified under a polariscope. Edges of the cut had a fire-polished appearance.

An example of cutting a modern, non-metal material is the cutting of KEVLAR-49*/EPOXY composite, Figure 10. Advanced composites have high specific strengths and modulii, but are very difficult to cut. In the prepreg form the semi-cured matrix gums up the cutting tool. In the cured state, special tools are needed to combat the abrasive action of the matrix and fibers. Machined edges of KEVLAR-49/EPOXY are fuzzy and structurally undesirable. Laser assisted cutting produces a sharp, clean edge with little or no discoloration in both the prepreg and cured forms, at speeds unrivalled by mechanical means. Similar success has been demonstrated on Boron/Epoxy composites which are normally cut with diamond impregnated saws.

* A product of EI Dupont de Nemours & Co. Inc.

CONCLUSION

Laser cutting of thin materials achieves the best quality and rate when the cutting occurs mainly by the vaporization of the material. The energy density required for such vaporization is easily provided by a TEM_{oo} Mode Laser. Based on data obtained with a 1000 watt TEM_{oo} Mode Laser, cutting rates for both metallic and non-metallic thin materials appear to be inversely proportional to the thickness of the material, while directly proportional to the input power beyond a certain threshold value. The laser assisted cutting rates achieved and presented in this paper indicate manufacturing cost effectiveness especially in the fabrication of hard-to-machine materials. With some variation in the cutting rates, it is with equal ease that Steel, Quartz, Aluminum or Kevlar-49 are cut. The increasing use of high strength alloys and composites, coupled with the spiraling costs of machine tools and labor, assures the future of power lasers in manufacturing.

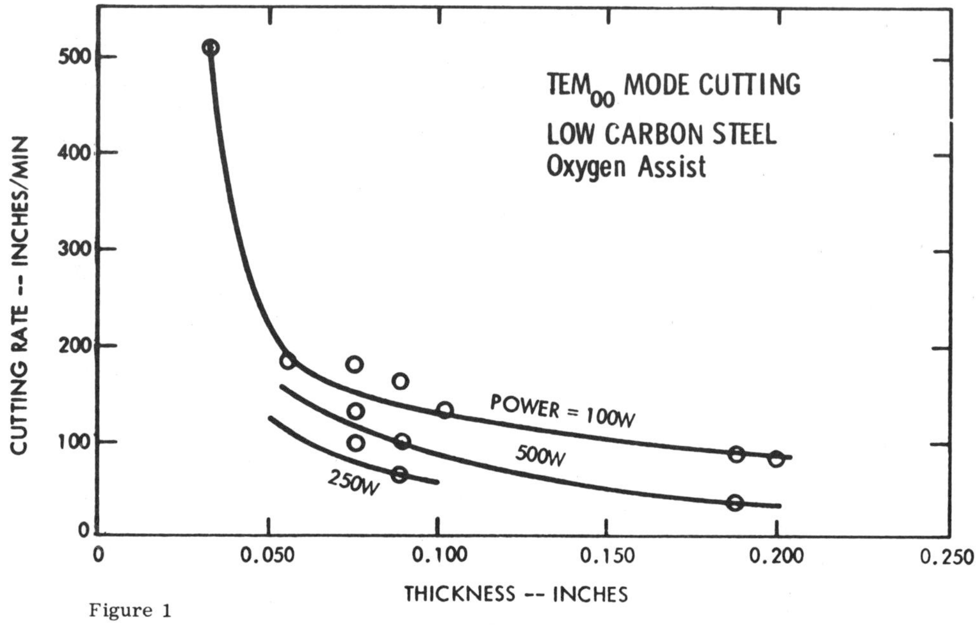

TEM_{00} MODE CUTTING
LOW CARBON STEEL
Oxygen Assist

CUTTING RATE -- INCHES/MIN
THICKNESS -- INCHES
POWER = 100W
500W
250W

Figure 1

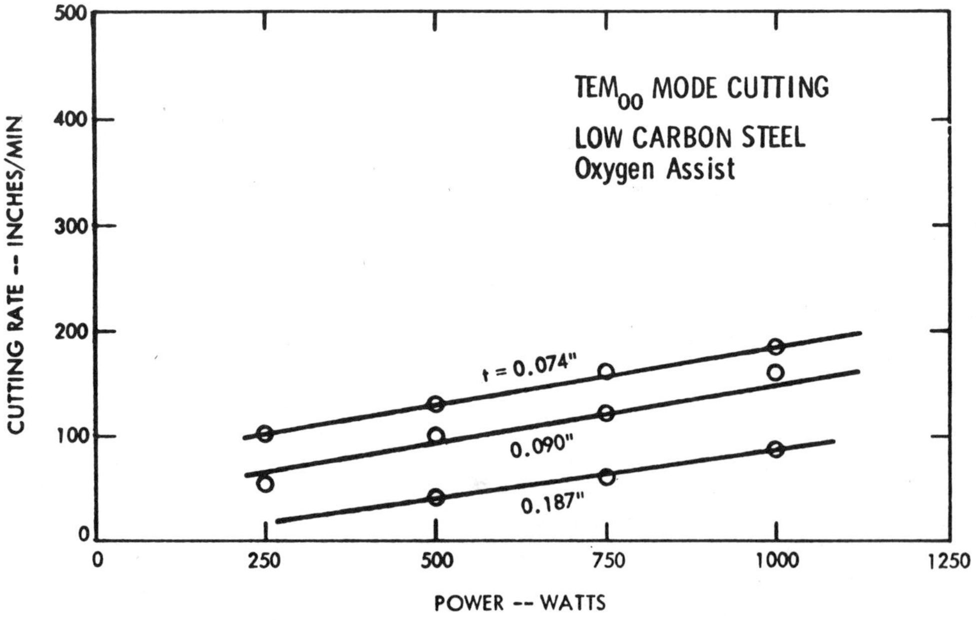

TEM_{00} MODE CUTTING
LOW CARBON STEEL
Oxygen Assist

CUTTING RATE -- INCHES/MIN
POWER -- WATTS
t = 0.074"
0.090"
0.187"

10/74

Figure 2

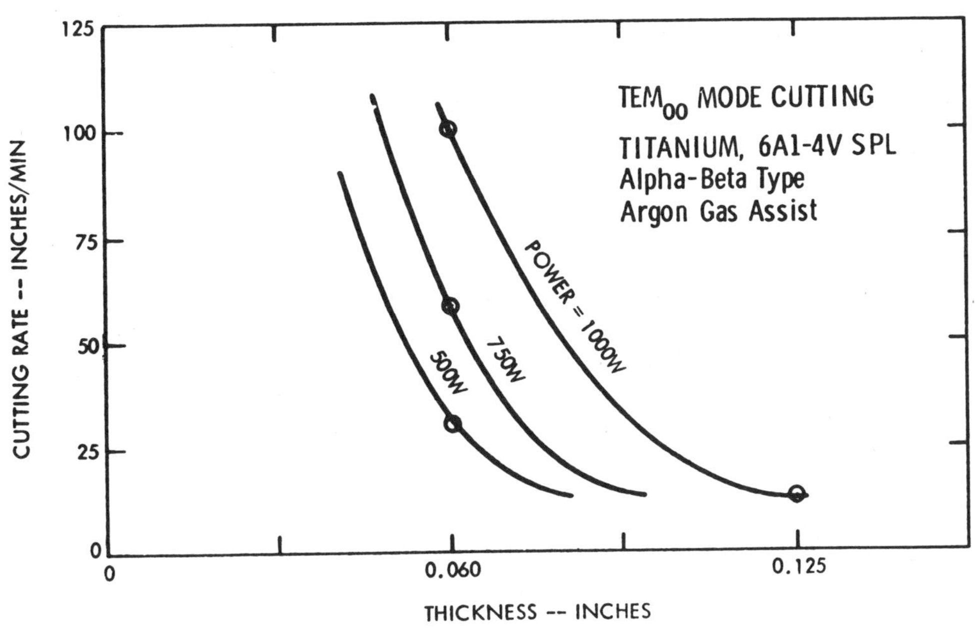

10/74

Figure 3

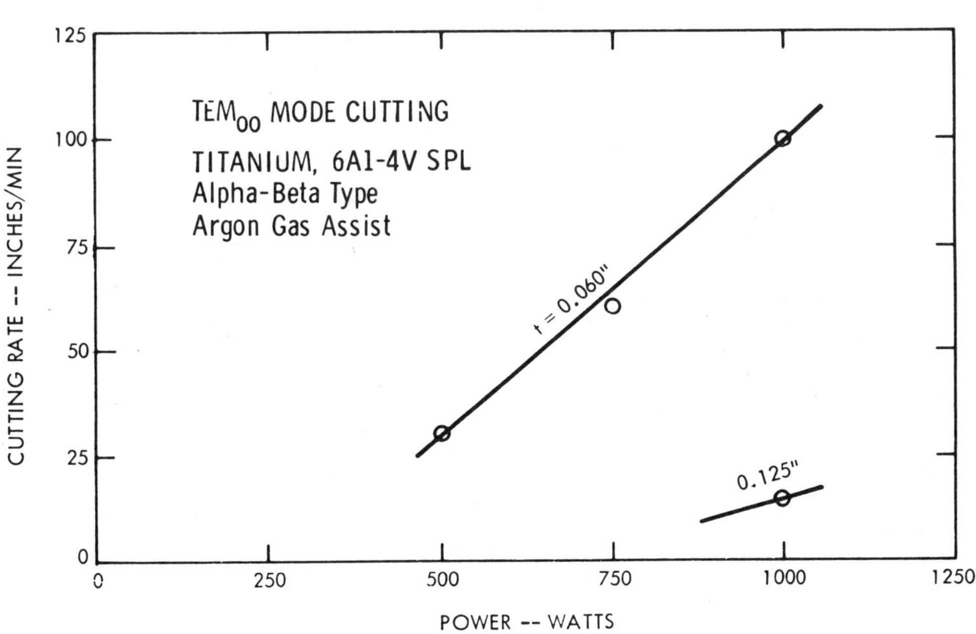

10/74

Figure 4

GTE SYLVANIA | ELECTRONIC SYSTEMS GROUP WESTERN DIVISION

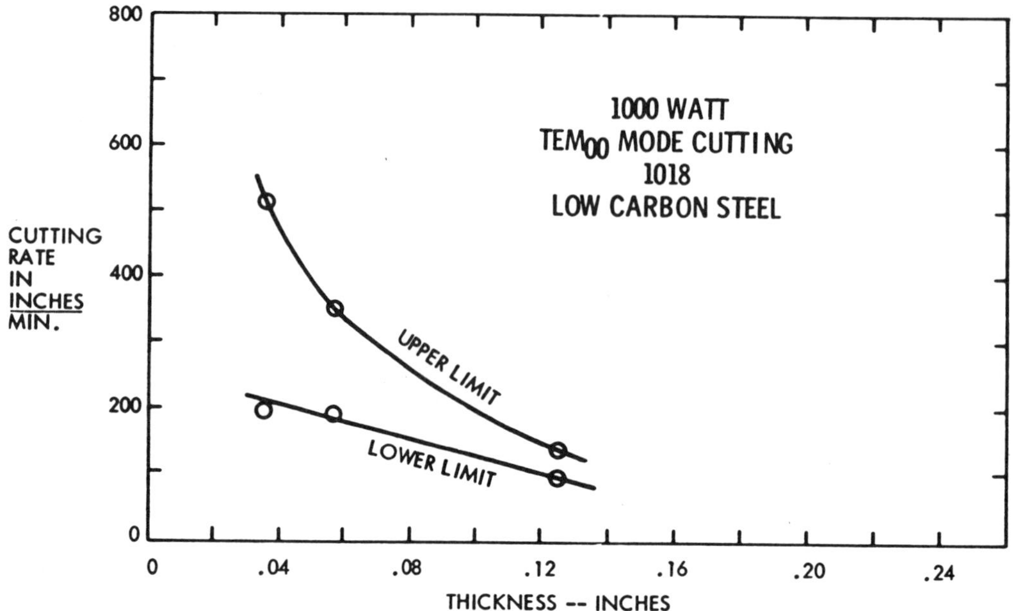

Figure 5

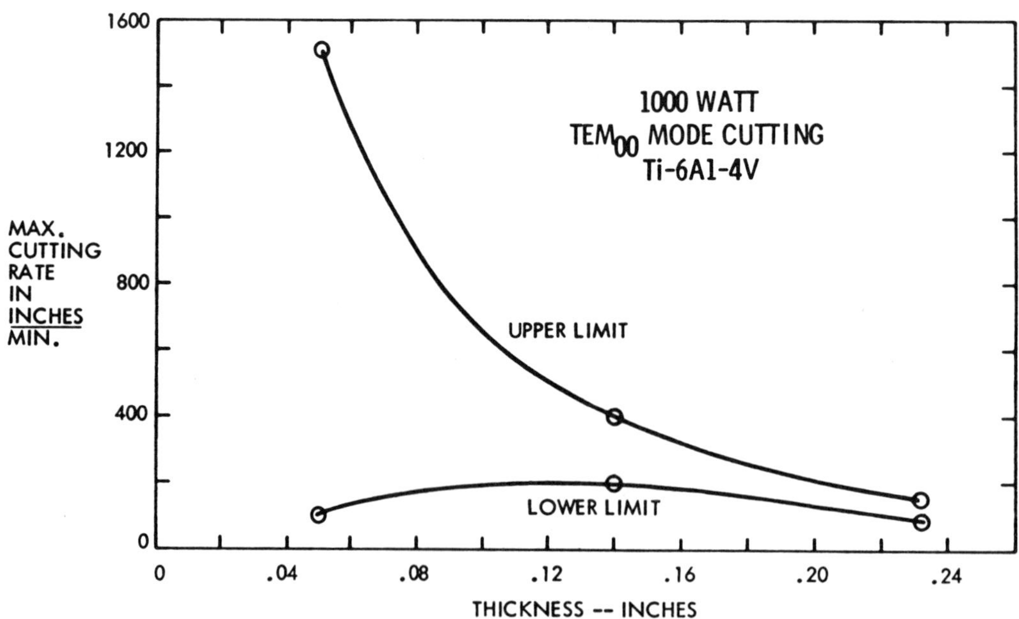

Figure 6

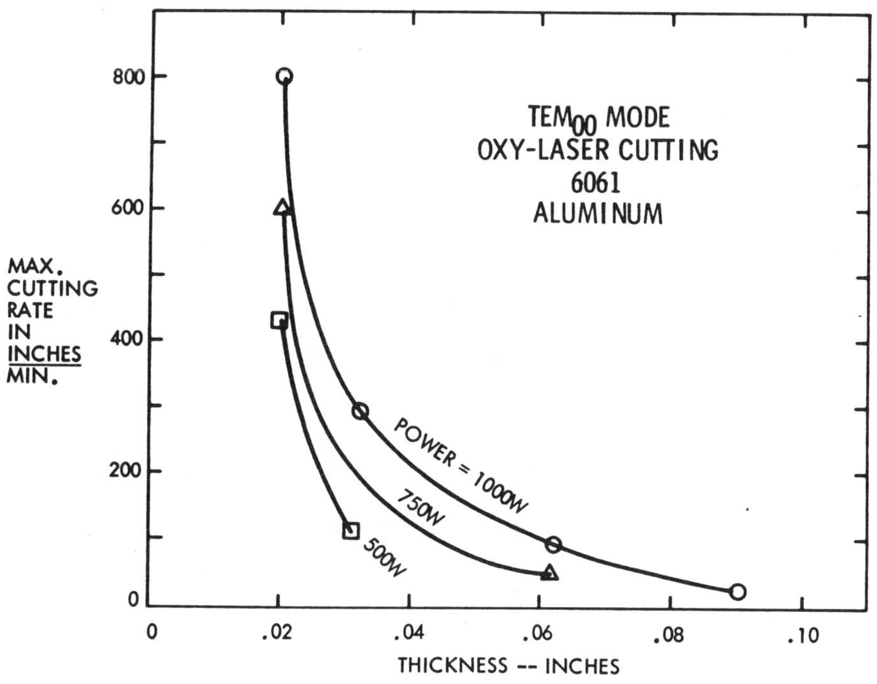

Figure 7

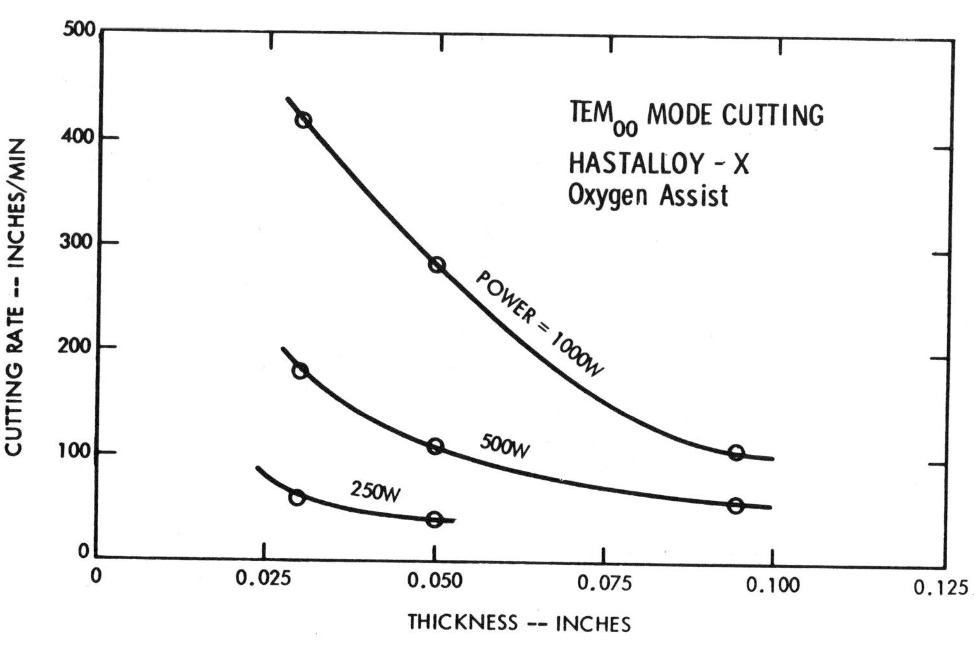

10/74

Figure 8

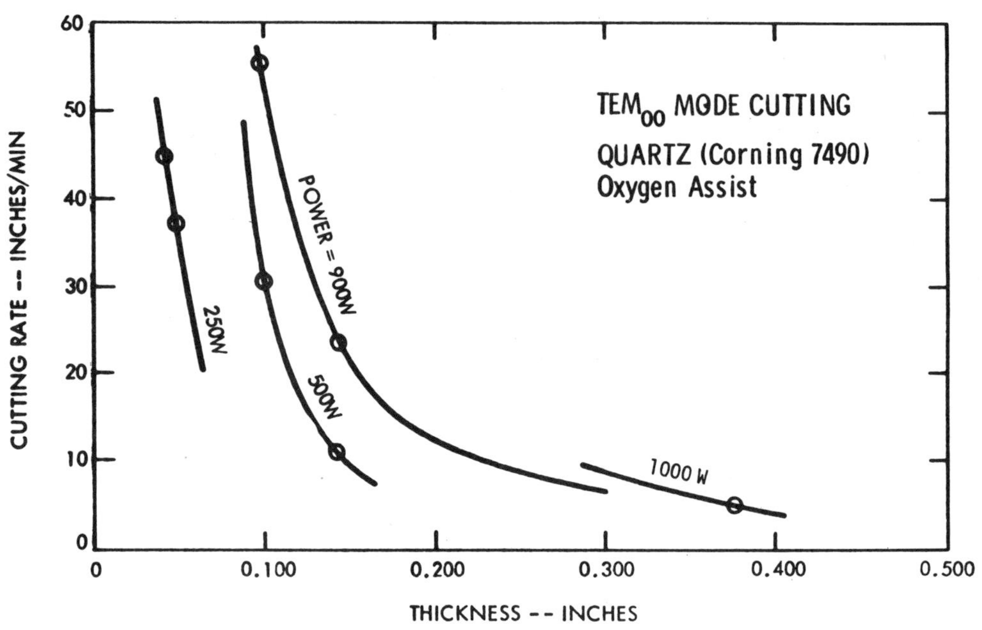

10/74

Figure 9

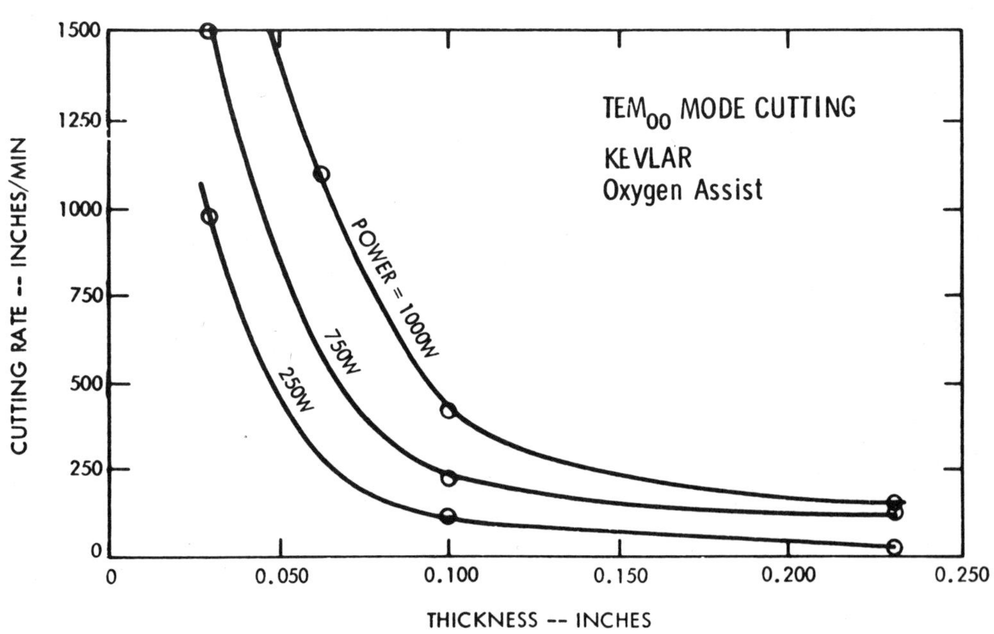

10/74

Figure 10

Reprinted from: Manufacturing Engineering, January 1976

Metal Machining with a Neodymium Laser

Normally limited to the drilling and machining of comparatively thin materials, the neodymium-YAG laser is proving effective with reactive materials up to 0.600 inch. Needed now: higher power density – and some attention from industry

CLARENCE A. PIPPIN
The Dow Chemical Co.
Golden, Colorado

SEARING A PATH through 0.199-inch thick titanium, at the rate of 39 inches per minute, this Nd-YAG laser may well be the forerunner of a new generation of machine tools.

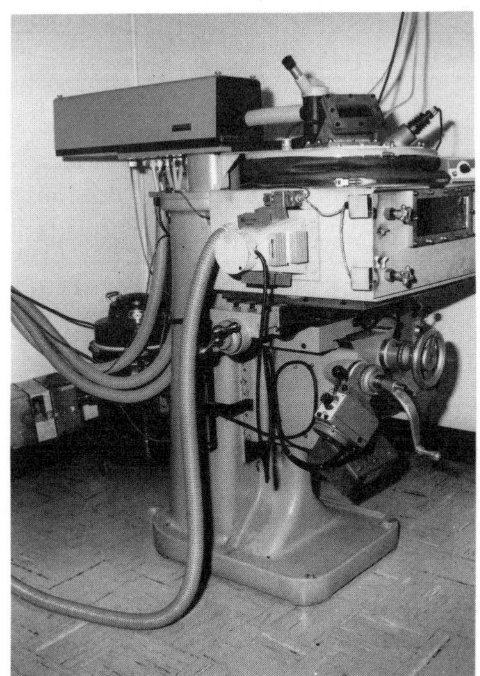

1. LASER MACHINE TOOL delivered to the Rocky Flats plant of the Atomic Energy Commission by Korad. Unit comprises a modified Korad laser and the base of a U.S. Burke milling machine.

Most of the work reported in the technical literature on metal machining by laser has been done using the CO_2 laser. Certainly, the CO_2 with its 10.6-micron wavelength is available in much larger sizes than any other laser. In general, the neodymium-YAG laser with a wavelength of 1.06 micron has been restricted to micromachining. Because the minimum focused area of the TEM_{00} mode is proportional to the wavelength squared, the neodymium-YAG laser could theoretically obtain the same power density as the CO_2 laser with only 1/100 of the power. Unfortunately, as the Nd-YAG laser power level is increased, higher order modes become dominant and this advantage is much reduced. The modes are denoted TEM_{mn} modes, where TEM stands for "transverse electromagnetic" and the subscripts *m* and *n* stand for the numbers of zeros in the beam intensity profile in the *X* and *Y* directions.

Most metals will absorb a higher percentage of the light at 1.06 micron than at the 10.6-micron wavelength of the CO_2 laser. Further, ordinary glass can be used as lens material for the 1.06-micron wavelength. These advantages led to research at the Rocky Flats Plant* of the Atomic Energy Commission (now ERDA) into the potential use of the Nd-YAG laser as a machine tool.

SYSTEM DESCRIPTION

An overall view of the laser machine tool is shown in *Figure* 1. The laser itself — at the top of the machine tool — is a Korad KY5 (modified) neodymium-YAG laser mounted on the base of a U. S. Burke vertical milling machine. A rotary table made by M & M Precision Systems, Inc., is mounted on the machine's worktable. This rotary table is driven by a variable speed motor.

Beam Configuration. The light comes from the laser head horizontally to strike an adjustable 45-degree mirror where it is reflected vertically downward. It then passes through a 75-mm

---

* Operated by The Dow Chemical Company until June 30, 1975, and then by Atomics International of Rockwell International. All development work on the project was performed under U.S. Atomic Energy Commission Contract AT(29-1)-1106.

focal length lens which focuses it to a small area for use. The lens is of ordinary optical glass, but is coated with an antireflective material for the 1.06-micron wavelength. A microscope type eyepiece and cross hairs allow accurate location of the workpiece. A gold-plated copper cone covers the bottom of the gas chamber. The gas in this chamber cools the focusing lens and also exits from the cone coaxially with the laser beam. This gas assists the drilling or cutting process. *Figure* 2 shows this cone and also the previously mentioned rotary table.

The design of the cone has not been maximized, but the use of a cone decreases the beam area by 60%. This is true in spite of the fact that the beam apparently does not come close to the walls of the cone, an effect that is not as yet understood. The laser beam exits from a 0.054-inch (1.37-mm) diameter hole in the tip of the cone as it approaches the "focal point." Actually, there is no focal point, but rather a zone in which the beam is essentially cylindrical for a distance of about 3/16 inch (4.76 mm) before it starts to diverge. The beam shape was determined with higher order modes present by drilling very thin stainless steel sheet at various positions perpendicular to the beam. The diameter of this cylindrical section as determined by this technique is about 0.009 inch (0.23 mm) which is approximately thirty times as large as the theoretical diameter for a Nd-YAG laser operating with TEM_{00} mode. This laser does not operate in the TEM_{00} mode at large power output levels.

Safety Requirements. In operating a powerful laser, certain safety precautions are necessary. A steel box 24 x 20 1/2 x 10 3/4 inches (61 x 52 x 27.3 cm) is mounted on the machine table to contain reflections. Three windows are provided in this box for visual observation. Each window contains two layers of plexiglass and a filter for 1.06 micron light. This 0.127-mm thick filter material is "Laser-Gard Film" made by the Glendale Optical Company. This box also protects the operator from any toxic metal fumes produced in machining. A negative pressure is maintained within the box and the fumes pass through an absolute filter on the way to the suction source. A full-width door on the front of the box contains one of the observation windows. Two microswitches in series are activated by closing the door on the box as an interlock to the shutter. This safeguard is necessary to insure that the laser light will not get into the room; the redundancy of switches provides protection against microswitch failure. The flexible boot between the moving and stationary parts of the box consists of an inner layer of silicone-impregnated asbestos cloth, two layers of aluminum foil, and an outer layer of billiard cloth.

Some laser investigators indicate that a Q-switched laser cannot be used for large metalworking tasks, but must be limited to micromachining problems. However, our research indicates that with the vibrating mirror type of Q-switch, this laser has considerable metalworking capability. The laser pulse is triangular in shape and the av-

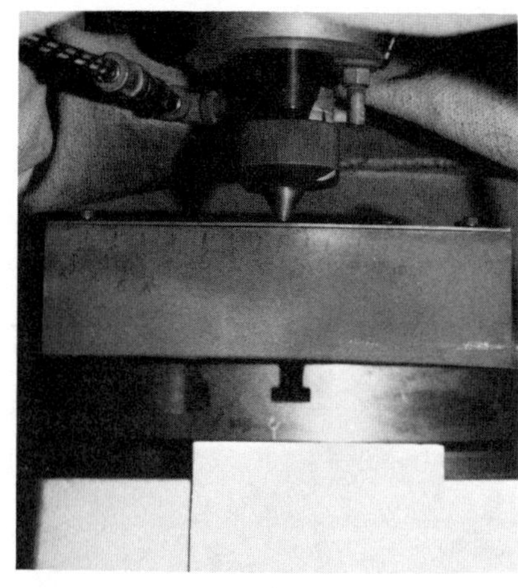

2. GAS CONE of the Nd-YAG laser. Gas cools the focusing lens and assists the machining process.

3. LASER CUTTING with an oxygen assist. From left to right are stainless steel, titanium, Zircaloy 2, and mild steel.

TIME REQUIREMENTS FOR DRILLING 0.004/0.005 INCH DIAMETER HOLES

Material	*Type 304 Stainless*	*Beryllium*	*Aluminum (3003)*	*Tungsten*	*Tantalum*	*Copper*	*Magnesium AZ31*	*Uranium*
Thickness	0.120 in. 3.05 mm	0.049 in. 1.24 mm	0.065 in. 1.65 mm	0.033 in. 0.84 mm	0.051 in. 1.30 mm	0.100 in. 2.54 mm	0.123 in. 3.12 mm	0.053 in. 1.35 mm
Drill time requirement (min.) with oxygen	1.46	4.63	Very long	0.32	0.44	1.63	Very long	0.20*
with air	1.33*	3.41	0.99	0.27*	0.35*	1.01*	12.50	0.22
with argon	3.69	0.56*	0.28*	1.34	1.88	Very long	0.36*	0.43
Lamp current used (amp.)	34	32	34	34	35	34	35	35
Average laser power of 1.06 micron wavelength (watt)	31	24	31	31	42	31	42	42
Optimum lamp current (amp.)	35	33	34	34	35	34	35	35
Maximum thickness drilled	0.190 in. 4.83 mm	0.198 in. 5.03 mm	0.122 in. 3.10 mm	0.100 in. 2.54 mm	0.105 in. 2.67 mm	0.125 in. 3.18 mm	0.123 in. 3.12 mm	0.125 in. 3.18 mm

* *Optimum gas for drilling*

erage length of each pulse is about 100 nanoseconds. The repetition rate is 4300 pps.

CUTTING METALS

There are two laser cavity configurations that are satisfactory for metal fabrication operations. The first configuration is called a close-coupled arrangement and only allows continuous wave (CW) operation to a maximum of 250 watts. This arrangement is used exclusively with the gas nozzle for cutting metals. The second configuration is normal coupling and allows Q-switched operation as well as CW at a lower level. This configuration was used in drilling holes.

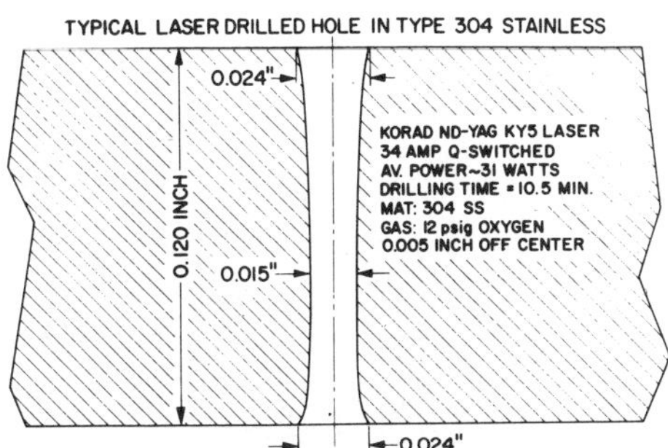

4. PROFILE of laser-drilled hole in Type 304 stainless steel.

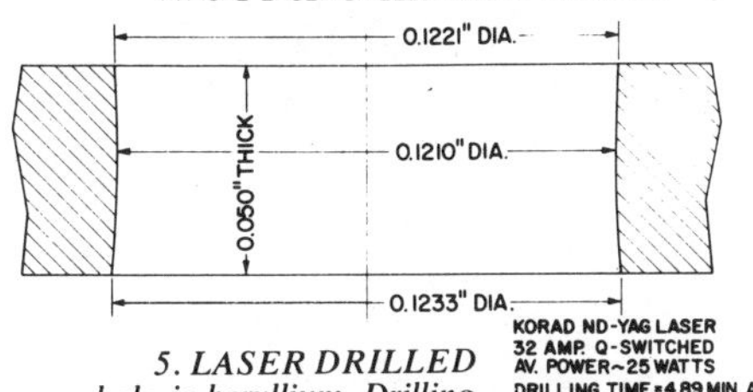

5. LASER DRILLED hole in beryllium. Drilling time is 4.89 minutes.

TYPICAL LASER DRILLED HOLE IN TUNGSTEN
0.014
0.010
0.091 INCH
KORAD ND-YAG KY5 LASER
31 AMP Q-SWITCHED
AV. POWER~23 WATTS
DRILLING TIME = 5 MIN.
GAS: 14 psig OXYGEN
0.005

6. SMALL HOLE drilled in tungsten. Uniformity of diameter is difficult to attain in extremely small holes.

In the CW mode, titanium 15.4 mm thick was cut at a speed of 990 mm per minute, which was the maximum servo speed. Oxygen was used at a pressure of 36 psig. Zircaloy 2 measuring 12.9-mm thick was also cut at 990 per minute, using oxygen at a pressure of 20 psig. Type 304 stainless steel, 1.5-mm thick, was cut at a speed of 585 mm per minute, using oxygen at a pressure of 30 psig. These cuts are shown in *Figure* 3. This picture also shows a cut of 3.0-mm mild steel. Although not shown here, it was demonstrated that when layered materials are cut, it is possible to cut the upper layer without damaging the underlying layer. This is true whether or not both layers are the same material. After determining some of the laser's capability as to cutting, the configuration was changed to give the Q-switched mode for drilling operations.

DRILLING METALS

In obtaining Q-switched operation, an orifice holder was substituted for the back mirror of the previous arrangement. The different sizes of orifices which can be inserted will determine the diameter of the beam allowed to go back to the vibrating mirror. For this work, a 0.250-inch (6.35-mm) diameter orifice was used. This size orifice does not noticeably reduce the power as compared to the maximum orifice, 8.33-mm diameter. In general, as the orifice diameter is decreased, the higher order modes are reduced. It is necessary to use a 1.42-mm diameter or smaller orifice to obtain a TEM_{00} or fundamental mode. In this event, the power output is quite low. When utilizing the 6.35-mm diameter orifice, this arrangement gives a maximum of approximately 100 watts continuous wave when the mirror is stationary and 50 watts Q-switched when the mirror is vibrating at its mechanical resonant frequency.

Roundness and Taper. The majority of the investigation on this laser concerned the drilling of holes. As indicated before, this laser does not give the TEM_{00} mode except at very low power levels, and therefore the hole drilled may not be perfectly round. In order to overcome this problem, the specimen was mounted on a rotary table and turned while the hole was drilled.

Taper is another problem associated with laser-drilled holes. It seemed that this problem might be corrected by tilting the beam slightly so as to give a larger diameter at the exit surface. The amount of tilt on this laser was 2.3 degrees in both the X and Y directions. This counteracted the laser's natural tendency to drill a tapered hole with the larger diameter being on the entrance side. Of course, the rotary table is turned during the entire drilling operation. Examples of the holes obtained are shown as *Figures* 4, 5, and 6. *Figure* 4 defines the hole profile of a 0.015-inch (0.38-mm) hole in Type 304 stainless steel 0.120-inch (3.04-mm) thick. *Figure* 5 defines the hole profile of a 0.121-inch (3.07-mm) hole in 0.050-inch (1.27-mm) beryllium. *Figure* 6 is a 0.005-inch (0.13-mm) hole in 0.091-inch (2.31-mm) thick tungsten. These profiles were determined by using a series of pin gages; penetration was measured with an optical comparator. In regard to *Figure* 6, it is likely that if tungsten were to be drilled again, a better hole profile would be possible.

Uniformity and Size. It is more difficult to drill a very small hole of uniform diameter in thick material than it is to drill a larger uniform diameter hole. (This does not refer to time requirements — only to diameter uniformity.) In some cases, it is accepted practice to move the focus position down as the hole is drilled. This technique was not used in any of the work discussed in this article, except for 0.250-inch (6.35-mm) uranium. In general, the moving of the beam focus as the hole is drilled appears to be desirable as thicker materials are used.

Type of material is also important. For instance as the metal thickness is

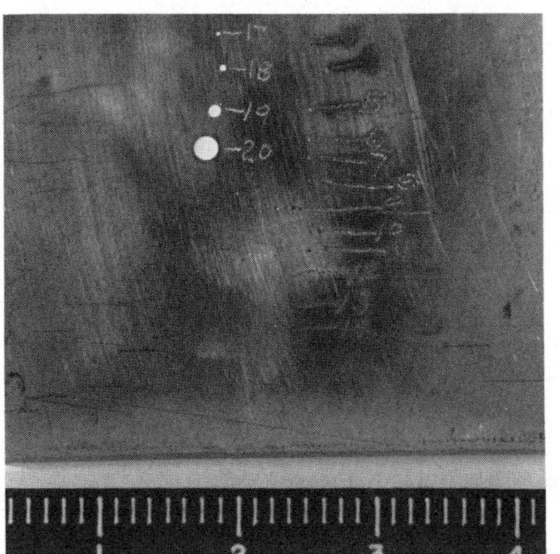

7. TYPE 304 STAINLESS drilled with laser. Workpiece sample is 0.117-inch thick. ▶

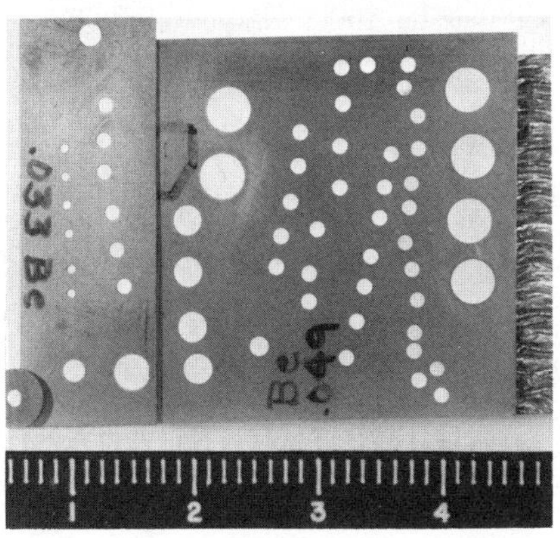

8. BERYLLIUM SAMPLES. Largest hole is 0.359 inch in diameter.

increased, the necessity of using this technique appears much sooner for beryllium than for stainless steel.

► STAINLESS STEEL. *Figure* 7 is a photograph of a 0.117-inch (2.97-mm) piece of Type 304 stainless steel. In general, the 14 small holes shown had a minimum diameter varying from 0.006 inch (0.15 mm) to 0.013 inch (0.33 mm). The inlet and outlet diameters of each hole were about the same and varied from 0.003 inch (0.08 mm) to 0.007 inch (0.18 mm) larger than the minimum diameter. Any size hole can be obtained depending upon the distance between the center of the beam and the rotary table's center of rotation. The minimum diameter for any hole is twice the off-center distance plus the diameter of the laser beam. The minimum diameter of a hole occurs approximately at the center of the piece for this thickness. The largest hole in this 0.117-inch (2.97-mm) stainless steel piece was 0.168 inch (4.27 mm) in diameter.

► BERYLLIUM. *Figure* 8 shows three pieces of beryllium. The 0.033-inch (0.84-mm) piece of beryllium has holes from 0.006 to 0.290 inch (0.15 to 7.37 mm) in diameter. The large hole is only 0.050 inch (1.27 mm) from the edge. The 0.049-inch (1.24-mm) piece of beryllium has six small holes of 0.003 and 0.004 inch (0.08 and 0.10 mm) in the marked-off section. A piece of No. 40 gage wire is threaded through two of the holes. The largest hole in this piece has a diameter of 0.359 inch (9.12 mm). The disc has a 0.127-inch (3.23-mm) diameter hole in it and is 0.071-inch (1.80-mm) thick. Not shown here is a hole of 0.500-inch (12.70-mm) diameter drilled in a piece of beryllium 0.118-inch (3.00-mm) thick. This required a special technique that cannot be disclosed at this time.

► MAGNESIUM ALLOY. *Figure* 9 is a photograph of a piece of AZ31 magnesium alloy which is 0.123-inch (3.12-mm) thick. The hole diameters vary from 0.005 inch (0.13 mm) to 0.166 inch (4.22 mm) for the largest hole. Beside this piece of magnesium is the center sections removed in drilling (trepanning) the two largest holes.

► TUNGSTEN. *Figure* 10 is a photograph of a piece of 0.033-inch (0.84-mm) tungsten and a disc of tungsten 0.060-inch (1.52-mm) thick. In the larger piece, numerous holes 0.004/0.005-inch (0.10/0.13-mm) diameter were drilled. In the thicker tungsten disc are holes of 0.003, 0.004, 0.005, and 0.121-inch (0.08, 0.102, 0.13, and 3.07-mm) diameter.

Power Levels. The various metals require different power levels for optimum operation. Likewise, the choice of gas in the gas cone depends on the metal being drilled. The accompanying table shows the operating conditions for eight metals, with the best gas for each metal indicated by an asterisk. Some metals must be protected by an inert gas (argon) while others are better helped by oxygen. Furthermore, in all cases investigated thus far — except for uranium — when oxygen was desirable, it was better to use air. Why oxygen in the air works better than pure oxygen has not been determined.

The maximum lamp current allowed for this laser is 42 amperes. When the laser is Q-switched, the average power output as indicated by an average reading power meter reduces at 40 amperes and higher. Also, even though the average power increases with lamp current up to and including 39 amperes, the drilling effectiveness does not. The table indicates that the most effective lamp current varies from 33 to 35 amperes depending on the metal being drilled. The maximum thickness of metal that has been drilled with a 0.003 or 0.004-inch (0.08 or 0.10-mm) diameter hole for the various materials are shown. In general, the limit for any metal is slightly under 0.200 inch (5.08 mm), but a 0.270-inch (6.86-mm) diameter hole was drilled (trepanned) through 0.250-inch (6.35-mm) uranium. A movement of the focal point was required for this depth of drilling.

The thicknesses of metal thus far drilled are not necessarily the limits for tungsten, tantalum, copper, and magnesium. However, it may not be possible to drill these metals to a thickness of 5 mm.

FUTURE POSSIBILITIES

The laser is too expensive a machine tool for use in ordinary machine shop operations. However, when there are unusual problems or difficulties that cannot be handled conventionally, the laser may supply the answers needed. These problems may include, but are not limited to:

1. Difficult materials such as tungsten or ceramics.
2. Necessity of drilling a very hard material.
3. Drilling a very small diameter hole in almost any material, especially if the depth exceeds 10 times the diameter. The maximum depth is normally considered to be 5 mm.
4. Drilling a small diameter hole in thin material or at an angle in thicker material where the tool pressure of an ordinary tool may cause the hole location to shift or the drill to break.
5. Drilling a hole near the edge of a brittle material such as beryllium where the edge is likely to break out.
6. Drilling a hole in a material in which machining damage may occur due to normal tool pressure. One such material is beryllium.
7. Cutting stainless steel when the oxy-acetylene torch will not.
8. Cutting only the outer layer of metal when there are two or more layers of metal present. This has been done when both layers are of the same material. Careful control is needed.

It is hoped that this report of the pioneer efforts toward the use of the neodymium-YAG laser in applications other than micromachining will encourage others to extend this work. Further, it is hoped that the manufacturers of neodymium-YAG lasers will be encouraged to extend the state of the art in order to increase the applications for this laser. Basically, the need is that of increasing the power density. ■

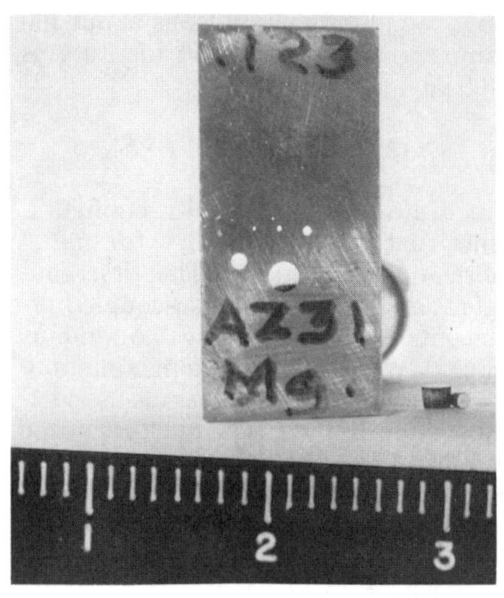

9. MAGNESIUM ALLOY containing holes ranging from 0.005 to 0.166 inch in diameter.

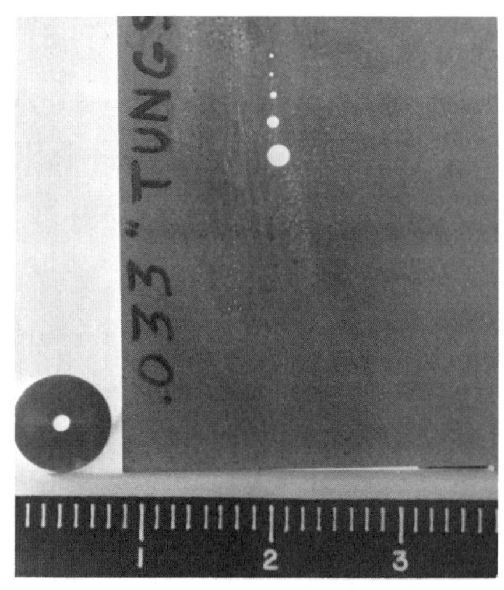

10. TWO TUNGSTEN SAMPLES. Disc at left is 0.060-inch thick. Plate – containing holes down to 0.004-inch diam. – is 0.033-inch thick.

Reprinted from: *European Plastics News, Janaury 1977*

Machining plastics by lasers

COMMERCIAL INDUSTRIAL LASERS are available in a range from a few watts to over 600 watts. For most plastics and rubber applications the smaller 50 watt and 150 watt models provide a machining capability unmatched by conventional techniques.

The laser produces an intense beam of infra red light which can be brought to focus onto a work-piece using simple lenses. Consequently, laser machining is a non-contacting technique ensuring not only a sterile machining process, but one without any tool wear or clogging.

Having focussed the laser beam to a very small spot, enormous power densities are generated which can melt or vapourize most materials. Because of the minute size of this focal point it is possible to drill a hole through plastics without affecting material in the immediate vicinity. Most plastics can be easily drilled by carbon dioxide lasers and throughput can be as high as 2 500 holes/second.

Two particular applications where lasers have excelled are the drilling of aerosol valves and the production of disposable medical products, such as catheters. Aerosol valves have normally been produced on injection moulding machines using pins in the mould to provide metering holes in the valves. However, with the introduction of laser hole drilling to this application, a number of major benefits are evident.

Key advantages are: reduction of inventory stock for valve manufacture by providing a drill-on-demand product; repeatability of product by not using moulds with worn or damaged pins; elimination of moulding pins, reducing initial mould cost; fewer rejects caused by broken or defective pins; standardized product with ability to introduce a variety of valve types.

Medical products too can benefit from laser machining where the sterile nature of the process is important. Not only are the parts drilled or slit in an hygenic manner, but the finished parts are free from debris, since laser machining ensures that all the material is vapourized by the sudden and intense pulse of energy.

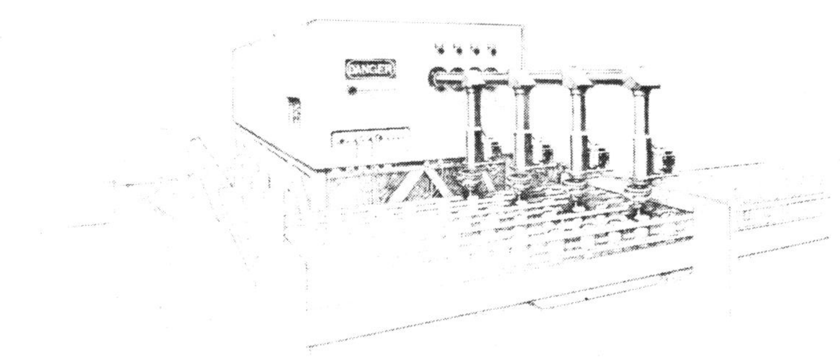

Schematic layout of industrial laser drilling application

One industrial laser handling system that can dramatically increase the cost efficiency of industrial laser applications has been developed by ZED Instruments of Twickenham. A major difficulty in fully exploiting the potential of industrial lasers has been the inability to present to the beam the parts or components with the speed and precision to match the laser's capability. ZED Instruments has used a combination of engineering expertise and optical know-how to overcome this problem.

A unique continuously rotated mirror assembly switches the laser beam into 5 time-sharing parallel channels. The mirrors and square motion 5 outlet lens assembly allow each channel to be pulsed up to 4 times with precision. The number of holes and hole size can be set by an unskilled operator while the machine is running. For maximum reliability the machine is mechanically driven and synchronised. The work carriers are positively indexed by a precision engineered cross-over cam and the helical and modified trapezoidal curve provides smooth intermittent motion at the high operating speed.

Typical drilling speeds are 1 200 holes/minute, but this can rise to 10 000 in suitable applications. Hole sizes may be as small as 0.15mm and as large as 0.5mm, and up to 20 different configurations may be programmed. Carefully designed interlocks protect the operator and the work, and the machine has been designed to operate in a normal industrial environment.

The ZED Instruments system incorporates a 150 watt Everlase CO_2 industrial laser supplied by Coherent Radiation of Bar Hill, Cambridge. This recently introduced laser has an exclusive gas recycler which reduces gas consumption and provides the lowest operating cost of any industrial laser available. The resonator structure, factory-locked optical prealignment and rugged cabinet enclosure ensure the reliability and stability essential for non-stop, round-the-clock production use. A laser unit of this type costs about £18 000 – although total equipment handling facilities can easily double this. Thus, to justify the expense, high volume and fast throughputs are needed, but running costs can be as low as £0.50/hour.

The latest system developed by ZED Instruments has been designed for drilling four holes in teats for baby bottles and in efficiency and output far supersedes the previous fairly crude methods. However, the versatility of lasers would also allow, for example, the cutting of polypropylene 0.5in thick at a rate of 10inches/minute (12.5mm at 254mm/min).

Whilst each application requires a fresh approach, and probably needs customer designed features, there are key factors in common for the cost effective and safe utilisation of lasers. One is beam kinematics, or a combination of optical, electronic and precision mechanical engineering for shaping and rapid positioning of the laser beam on to the product to be machined. For example, the teat perforating machine delivers a perfectly focussed beam in 20 well defined stationary positions in less than 0.5 seconds.

Machine kinematics, or the fast, accurate and reliable placement of the product into the work station is essential and ZED Instruments has considerable experience in high speed part feeding, indexing and continuous-motion part handling and positioning. The equipment is designed to give full protection to the operator, to the product, and to the machine itself. Efficient fume extraction safeguards the optical parts and protects the environment.

Finally, total systems concept involving beam handling, product handling and safety, together with the user's own technology, must be treated as interacting parts of the same system.

Industrial laser applications for plastics and rubber

	Material/Thickness	*Operation*	*Speed*
1	PP, 12.5mm	Cutting	254mm/min
2	PP, 0.38mm	Drilling	381mm/min
3	Contact lenses, 0.25mm	Drilling	5msec/hole
4	Rubber valve, 2.3mm	Drilling	0.5sec/valve
5	Silicone catheters, 3.05mm	Drilling	360msec/hole
6	Polymer mesh, 0.23mm	Cutting	1 676mm/sec
7	Acrylic, 12.5mm	Cutting	152mm/sec
8	Aerosol valves, 3.1mm	Drilling	5-20msec
9	Polyester film, 0.25mm	Drilling	711mm/sec
10	Silicone tubing, 2.5mm	Cutting	2sec/cut
11	Aerosol valves, 0.64mm	Drilling	5-20msec

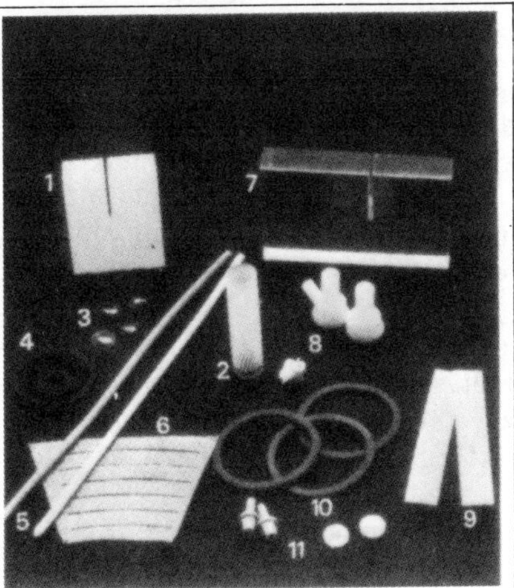

Reprinted from: Laser Focus, June 1976

The news in focus

Making tools more cost-effective

An LF Meeting Report, by Jean-Claude Diels

PROSPECTIVE INDUSTRIAL CUSTOMERS need two technological improvements in laser systems: time-sharing, to permit continuous use of a single laserbeam at widely separated sites; and a beam-direction system based on fiber or waveguide optics, to replace bulky articulated arms and their enclosed optics.

Lasers are fast and precise, but time-sharing and improved beam delivery would help to justify their cost

These conclusions emerged from the three sessions on laser machinetools for nonlaser specialists conducted Mar. 10 during Westec '76 at the Los Angeles Convention and Exhibition Center. The laser sessions clearly left the attentive audience of about 100 engineers with the impression that lasers could be highly efficient heat sources in certain cutting, welding and surface-treating applications. Equally evident, however, was the need to make a careful economic analysis of competing nonlaser methods because of the high costs of the large lasers described by makers and users of industrial systems.

The laser sessions were organized by Stephen M. Copley of the University of Southern California's material-science department, Michael Bass of USC's center for laser studies and Kenneth Kinyon, deputy director of advanced manufacturing systems and technology at the McDonnell Douglas Aircraft Corp. Westec was sponsored by the American Society for Metals of Metal Park, Ohio and the Society of Manufacturing Engineers.

1 Point-source cutting makes lasers valuable in aerospace industry

The major applications of drilling and cutting with carbon-dioxide lasers are in the aeronautical industry. John Huber, group head of electromechanical advanced development at the Grumman Aerospace Corp. described an oxygen-assisted laser cutter which focuses the 250-watt CO_2 beam with a 2.5-inch diameter lens through a nozzle blowing gas onto the workpiece. Since a high-quality cut requires accurate vertical positioning of the nozzle and a converging gas flow into the cutting region, sensors and automatic controls keep the nozzle 0.020 ±0.005 inch above the workpiece. To keep the cut straight, the nozzle was designed to provide a helium flow surrounding a laminar flow of oxygen. With such a nozzle design, the depth of the heat-affected zone is less than 0.010 inch at feed rates exceeding 2,000 inches per minute of 0.050-inch titanium-alloy sheets. Rather than moving the workpiece, the nozzle was put on a digitally controlled articulated arm.

Since there is no cutting force with a laser, positioning of the workpiece is easy and does not require a rigid clamp. An additional advantage over the bandsaw is that the laser is a point-source cutting system; a cut can be initiated at any location, since the laser also does the drilling. The combination of minimal material upsetting and preparation, no tool material contact and point-source cutting with laser cutting provide savings in labor as well as making possible optimal "nesting" of the pattern to be cut, thereby saving material. In current applications, savings are 75% for cutting F-14A stabilizer plates and wing stringers of titanium, 70% for EA 68 doublers of 4340 steel and 80% for E-2C nacelle bulkheads of 321 stainless steel. The absence of tool-material contact is essential in cutting such thin foils as 0.002-inch prepressed band.

While it is relatively easy to assign a constant speed to a digitally controlled nozzle, the range of optimal cutting speed generally is broad enough for manual operation. John Williamson, project manager for the Air Force Materials Laboratory at Wright-Patterson AFB, described a multiaxis, hand-operated, CO_2 laser cutter. Here, the critical distance between lens and workpiece is set by mechanical contact; alignment to the workpiece and cutting speed are a function of operator skills. The full unit, including vacuum pump and power supply, is mounted on a wheeled platform and can be moved conveniently.

Cutting aluminum with CO_2

While the economy achieved by laser cutting is obvious for titanium, laser cutting is more difficult for aluminum, which makes up at least 75% of any airplane. Birger O. Anderson, general supervisor of manufacturing research and development at the Boeing Co., expects the advantages of thin cut, a cutting speed of 100 to 700 inches per minute, no fixtures, and no post-processing, to outweigh the high costs of a six-kilowatt CO_2 system being studied at Boeing.

The high costs are a major drawback in laser applications, and make time-sharing seem essential for economical use of highpower CO_2 lasers. In conventional systems, the laserbeam is dumped between operations; time-sharing would permit continuous use of the beam, speeding amortization of capital costs and reducing operational costs. Williamson proposes a central laser with mirrors to direct the beam towards the articulated arms of various cutting heads.

Another improvement, suggested by Michael Bass, associate director of the center for laser studies at USC, is a flexible waveguide to replace the bulky tubes and articulated arms used to direct the beams. Such a flexible metallic waveguide for CO_2 laser radiation is being developed at USC [see *News in Focus*]. A one-meter, lowloss prototype has been demonstrated, and versions for use with high-power CO_2 lasers can be forseen in the next decade.

2 Solidstate-laser welding requires pulses longer than for drilling

While drilling with pulsed solidstate lasers requires pulses as short as a fraction of a microsecond the ideal welding pulse must last 10 milliseconds, with an initial spike reaching 10^6 watts per square centimeter and the remaining pulse nearly constant at 10^5 w/cm^2. Once material being welded has reached the melting temperature, it is essential to limit deposition of heat by the laser radiation to avoid vaporizing the metal; in contrast the aim of drilling is to vaporize all the metal in the beampath. Longer focal lengths can be used for welding at the 1.06-μm wavelength of neodymium lasers than for drilling and cutting at 10.6 μm because smaller power densities are required for welding and the diffraction losses are lower at 1.06 μm, so height of the lens above the workpiece is not critical for the shorter wavelength.

As emphasized by Eugene Maloney, laser-products manager at the Raytheon Corp., and Michael Weiner of the Korad division of Hadron Inc., the advantages of laser welding include

- localized heating
- ability to join dissimilar metals
- easy fixturing of the pieces to weld, because no force is applied to them
- no surface separation; the beam itself can strip the insulation from electrical wires
- metallurgical control
- realtime verification of welds by observing and analyzing the plume of vaporized metal.

Weiner described one application for which the laser is ideal: thermocouple welding inside control rods of nuclear reactors. At Korad, such welding is done at right angles to the laser axis by deflecting the beam shortly before the focal spot as shown in the accompanying diagram.

Reducing capital and operating costs

The high acquisition cost of a laser system remains the major barrier to acceptance. To avoid the need for heavy and expensive highvoltage capacitors, Raytheon energizes flashlamps by half-cycle rectification of three successive phases of the main sector in its 40-w and 150-w SS380 and SS480 laser welders, with repetition rate of 10 hertz.

In addition to the need for speeding amortization of the laser's cost, time-sharing of the laserbeam is dictated by the requirement that the laser remain in operation between welds to ensure stable output. A two-stage time-sharing arrangement is contained in Raytheon's SS400 welder, with which the Bell Telephone System welds as many as 130 wire terminals on a small shunt plate for an electronic telephone-switching system; polyurethane coatings are removed from the wires by the initial part of the pulse. To minimize periods while the laserbeam is not in use, the welder employs two 10-inch-square x-y stages; one stage is loaded or unloaded while a minicomputer directs the other stage through its programmed motion at rates of four inches per second under the 0.25-inch-diameter focal spot. A servo system directs the laser to fire as the target moves under the laserbeam; pulse rates are 6 to 8 per second. One such system is in operation at a Western Electric Co. facility in Columbus, Ohio; a second system has been delivered to a Western Electric plant in Oklahoma City.

Jean-Claude Diels *is a research scientist at the University of Southern California's Center for Laser Studies*

Design And Development Of Equipment For Laser Wire Stripping

By W. F. Iceland
Project Engineer, Rockwell International, Space Division

For more than a year, the Space Division of Rockwell International has been laser-stripping single-conductor, Kapton-insulated wire, such as that used in the Space Shuttle orbiter.* To perform this work, the Division designed both bench-model and hand-held laser wire strippers. In this stripping process, an optical-mechanical system first rotates the focus spot around the wire, making a circumferential strip. The focus beam is then translated axially to slit the slug, which can be easily removed by hand. While Kapton is highly absorptive of the 10.6-μm and 1.06-μm laser wavelengths, the nickel coating of the copper wire is highly reflective, and the residual heat input is rapidly conducted away. Thus the integrity of the conductor itself is not compromised.

INTRODUCTION

Kapton-insulated wire, which is used in the Space Shuttle orbiter, has a very high melting temperature. This greatly limits the application of current wire-stripping techniques to such wire: thermal stripping methods are slow for production-type operations, while mechanical stripping, presently baselined for orbiter production, requires frequent calibration because of mechanical wear or operator tool abuse.

Searching for a more efficient stripping method, the Space Division designed and built three laser wire strippers. Two of these incorporate a CO_2 laser (10.6 μm); the third stripper employes a 1.06-μm Nd-YAG (neodymium-yttrium-aluminum-garnet) laser. Two of the lasers are operational in the Division's manufacturing wiring area.

The bench-model stripper, in production use over a year, is powered with a CO_2 continuous laser. This laser, built by Coherent Radiation Laboratories, is capable of producing a power output of 250 watts, but, for this application, its power output is limited to 15 watts. The laser is assembled with the Rockwell-designed, electronically controlled optical-mechanical stripping tool. Motion is provided by

*Under contract with the National Aeronautics and Space Administration (NASA), the Space Division is producing the Space Shuttle orbiter and serving as integrator of the entire Shuttle system.

electric motors with servo speed controls. The focused laser beam is rotated around the stationary wire, then axially along the wire. The Space Division has developed interchangeable collets for stripping wire sizes No. 26 through 1/0 (Figure 1).

A hand-held prototype laser stripper, which recently was installed in the production area at the Space Division, is powered by an Nd-YAG laser with a continuous-wave output of 5 to 7 watts. The laser assembly is mounted on a cart with power supply and cooling unit. The laser beam is transmitted through a dope-deposited, fused-silica-fiber cable to a movable stripping mechanism supported by a counterbalanced arm. The assembly has an auxillary low-power HeNe laser, which produces a red light for alignment. The optical-mechanical system, smaller than that in the bench model, has the same wire-stripping capacity: No. 26 through 1/0 AWG.

A second hand-held stripper, with a five-inch-long Hughes CO_2 laser inside the stripping head, is under development at the Space Division. The stripper mechanism, held in the hand and supported by a counter-balanced arm, will be tethered to a cart by the power cable and cooling hoses. The laser output will be 1 to 2 watts, and the unit will be capable of stripping Kapton-insulated wire sizes No. 26 through 14.

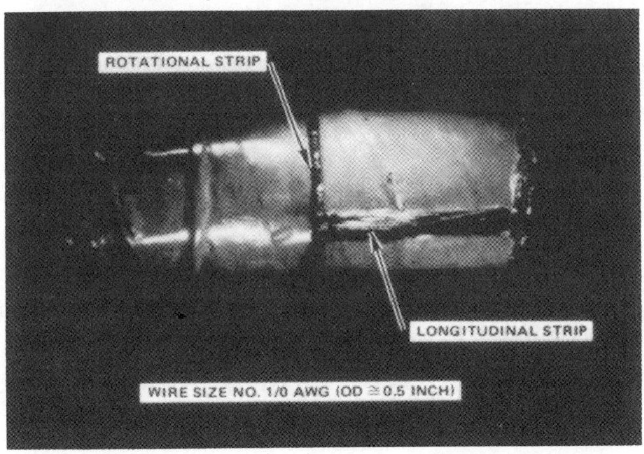

Figure 1. Laser-Stripped Wire

LASER CHARACTERISTICS

The Space Division's Advanced Manufacturing Technology Group developed the laser wire-stripping process. By using a laser in conjunction with a focusing lens, the investigators were able to create a concentrated heat source. This concentrated energy, which was focused on the rotating wire surface, rapidly melted or vaporized the insulation without adversely affecting the copper wire or its plating material. A stream of gas (air, inert gas, etc.) that was also directed onto the rotating wire effectively removed the insulation residue.

SYSTEM DEVELOPMENT

One of the early developments in the Space Division's laser wire-stripping process was the tool that rotates the wire under the laser beam. As shown in Figure 2, the wire is inserted into the spring-loaded vise and simultaneously positioned so that it is directly under the focused laser beam. The beam is then directed onto the wire surface as the laser shutter is actuated and the wire is rotated and moved axially by a geared, dc, permanent-magnet motor.

Using a Coherent Radiation Laboratories Model 41 laser, modified to operate in the 0.5- to 15-watt range, the investigators conducted a number of evaluations. In addition to power requirements, several lenses were evaluated to determine the optimum focal length.

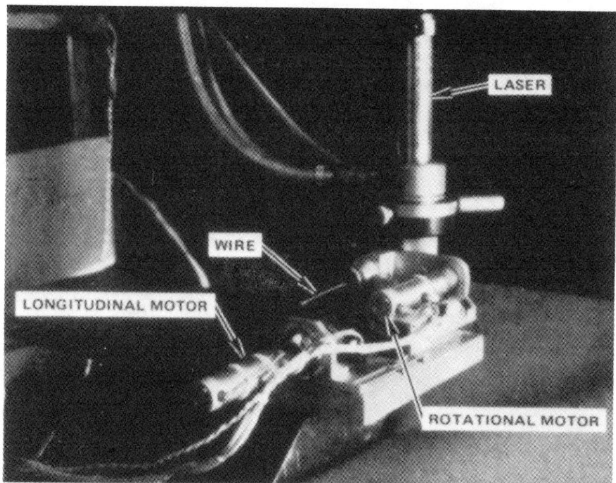

Figure 2. Rotational and Longitudinal Motion Control

The lens is basically an optical device that collects the laser beam and concentrates it on a small area (Figure 3). The smaller the area, the greater the energy or power density at the point of focus. For lenses with low-distortion optical properties, the diameter of the focused spot is given by

$$d_d = (10^{-4}) \frac{\lambda f}{D} \qquad (1)$$

where

f = focal length (FL) of the lens (inches)

d_d = diameter of spot (inches)

λ = wavelength of laser beam (10.6 microns for CO_2 laser, 1.06 microns for Nd-YAG laser)

D = diameter of lens aperture (inches)

To calculate power density, the laser power is divided by the area of the focused laser beam impinging on the Kapton wire surface:

$$P_D = \frac{W}{A} \qquad (2)$$

where

A = area of spot (square inches)

P_D = power density (watts per square inch)

W = laser power (watts)

Another series of tests was conducted to evaluate the gas-jet assist. Three candidate gases were evaluated: oxygen, argon, and compressed air. Figure 4 is a diagram of the basic gas-jet-assist laser wire-stripping system used in conjunction with the focused laser beam. The beam is reflected down through the focusing lens, through the chamber that forms the gas nozzle, and then focused onto the Kapton-insulated wire. The beam melts or vaporizes the wire insulation, while the gas jet expels or disperses the molten reaction products.

Besides removing the vaporized material, the gas jet serves the secondary function of maintaining a positive pressure in the gas chamber, thereby preventing undesirable debris from forming on the lens. An additional exothermic, or burning, action is achieved when oxygen is used as the gas jet.

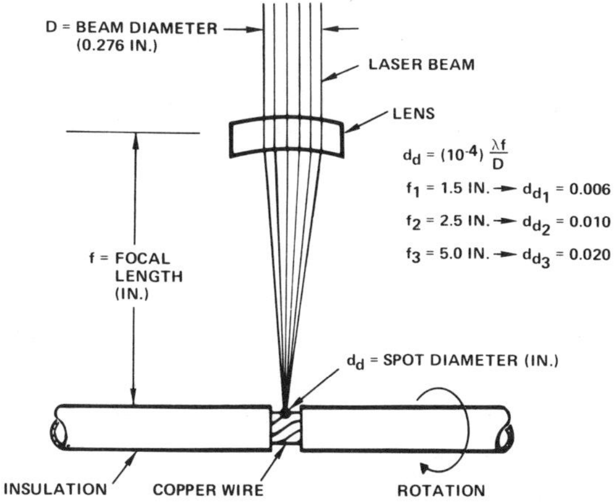

Figure 3. Optical System for Laser Wire Stripping

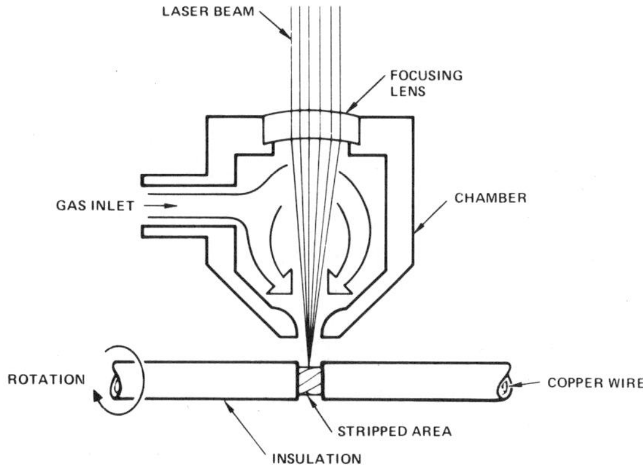

Figure 4. Laser Gas-Jet Assist

RESULTS OF EVALUATION

Table 1 summarizes the time required to strip Kapton-insulated wire of various gauges. At the 1-watt level, the minimum rotational stripping time is 2.58 seconds, which exceeds the 1.0- to 1.5-second requirement. Satisfactory stripping times were achieved at the 3-watt level for both 1.5- and 2.5-inch focal length lenses. It should be noted that an increase in laser power increases the stripping speed. Below 3 watts, changing from oxygen to argon or compressed air for the gas-jet assist is unsatisfactory (stripping time increased), and the argon stripping action is inconsistent.

At the 4-watt level, all combinations in Table 1 are acceptable. However, the minimum power level that results in the narrowest strip with the least vapor deposition on the conductor is 3 watts. Laser stripping at 3 watts, using a 1.5-inch focal length lens and oxygen, optimizes this process for No. 18 AWG wire.

Table 1. Total Wire-Stripping Time

	Time (seconds)*						
Power (watts)	1.5-in. FL Oxygen AWG No. 18	2.5-in. FL Oxygen AWG No. 18	5.0-in. FL Oxygen AWG No. 18	1.5-in. FL Argon AWG No. 18	1.5-in. FL Air AWG No. 14	1.5-in. FL Oxygen AWG No. 24	1.5-in. FL Oxygen AWG No. 14
1	2.58	4.25	10.63	**	31.25	13.25	–
2	1.46	–	–	**	8.5	3.85	1.66
3	1.18	1.58	2.4	–	1.29	1.55	0.89
4	0.99	1.21	1.46	1.04	–	1.5	0.5
5	–	–	–	0.73	–	–	–

*One revolution plus 3/16-inch longitudinal travel.
**Stripping inconsistent below 3 watts.

PRODUCTION BENCH-MODEL LASER WIRE STRIPPER

After Space Division and NASA personnel had reviewed the tabulated data, the Space Division was authorized by NASA to proceed with the development of a bench-model laser wire-stripping unit (10.6-μm wavelength). To accommodate harness wires, a design concept was selected in which the focused laser beam is rotated around the fixed wire (Figure 5). As shown in Figure 5, the coherent laser is initially reflected by a fixed mirror, then passed into the rotating joint and through the movable focusing optics. The three movable mirrors and output focusing lens optically rotate the laser beam, deflecting it onto the wire surface. Linear, or axial, travel of the mechanism moves the beam along the wire.

Figure 6 illustrates the working mechanism of the bench model. The physical design requirements limit the mirror surfaces to a diameter of 0.2 inch. To accommodate this dimension, the diameter of the laser

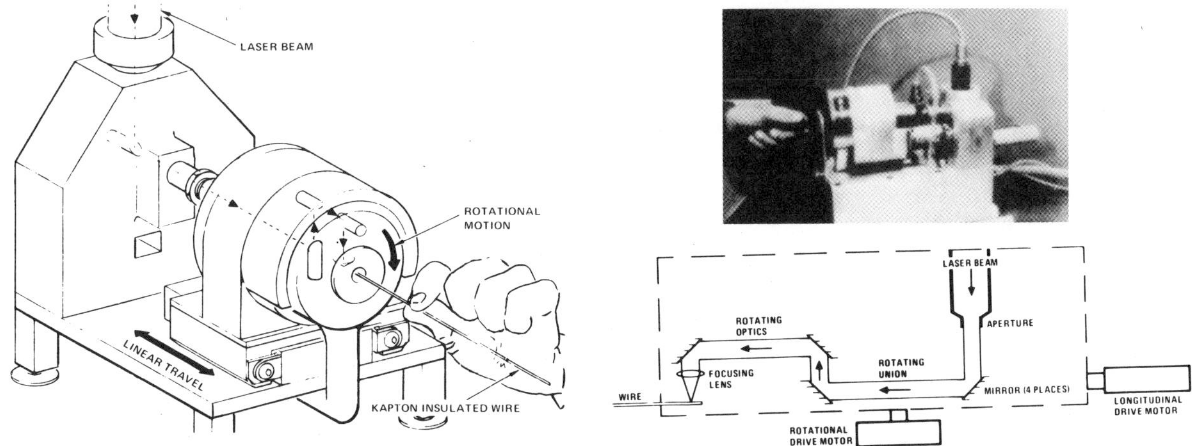

Figure 5. Design Concept for Laser Wire Stripper

Figure 6. Test Setup for Laser Wire Stripping

beam is reduced by the aperture. An output focusing lens focuses the rotating beam to a spot size of 0.005 inch. A laser power density of 10^5 watts per square inch enables the unit to strip wire ranging in size from No. 26 to 1/0.

Figure 7 shows the completed laser wire stripper as it appears in the Space Division's wire-stripping facility. The laser beam, generated in the horizontally mounted laser, is deflected downward by a 45-degree mirror, through the aperture, and into the rotating mirror-lens system. The wire is inserted into the fixture through a swinging frame, which can be moved during installation of the wire collet into the mechanism. A set of collets, one for each wire size, is shown in the photograph.

The stripping process is initiated by inserting the wire into the collet, which positions it directly in line with the laser beam. A shutter is actuated, and the focused laser beam impinges directly on the wire surface. The beam is rotated by a dc, permanent-magnet motor. The latter function is synchronized by a relay-limit switch. A geared, dc motor with feedback generator is used in conjunction with a dc amplifier to control velocity. This control system is designed for a 50 to 1 speed range, with a maximum rotation speed of 240 rpm's.

Oxygen, inserted through the rotating joint, is used as the gas-jet assist. By its oxidizing (exothermic) action on the insulation, oxygen helps the laser beam achieve a quality strip. A vacuum system removes debris and vapors formed by the stripping action, thus ensuring a safe working area.

In designing this system, the Space Division met all safety requirements dictated by Rockwell International specifications. Moreover, an eight-hour training period is required to qualify an operator to run a production laser wire stripper.

HAND-HELD LASER WIRE STRIPPER

In response to a request by NASA's Johnson Space Center, the Space Division instituted a program to develop a hand-held laser stripper. Figure 8 illustrates this device, a portable laser wire stripper capable of stripping individual Kapton-insulated conductors in Space Shuttle wire harnesses (No. 26 through 1/0).

This Nd-YAG laser system (1.06-μm wavelength) consists of a laser, fiber optic bundle, focusing optics, and electronic-sequencing motorized controls. Design of the system is similar to that of the bench model: a focused laser beam rotating around the fixed wire. As shown in Figure 9, the beam is processed through the fiber optic assembly, then through the beam-collimating lens (which recollimates it for the 0.2-inch-diameter rotating mirrors), and focused to a small-diameter spot by the output focusing lens. This design ensures that power density is available to strip wire sizes No. 26 through 1/0.

The Nd-YAG laser and fiber optic assembly were designed by the Coherent Radiation Laboratories. The 12-foot fiber optic assembly, which has a total efficiency of 85 percent, delivers a beam whose spot size is 70 microns at a full-angle beam divergence of 0.28 radians. At the exit end of the fiber, the power output is between 5 and 7 watts, multimode, continuous wave (CW).

Recollimation reduces the beam divergence from 0.28 radians to 2 milliradians. Since the focused spot size is the product of the beam divergence and focal length of the focusing lens, a short focal length lens causes the beam to strike the wire surface at a spot with a diameter of approximately 0.003 inch. With this spot size, power density greater than the minimum required (10^5 watts per square inch) becomes available.

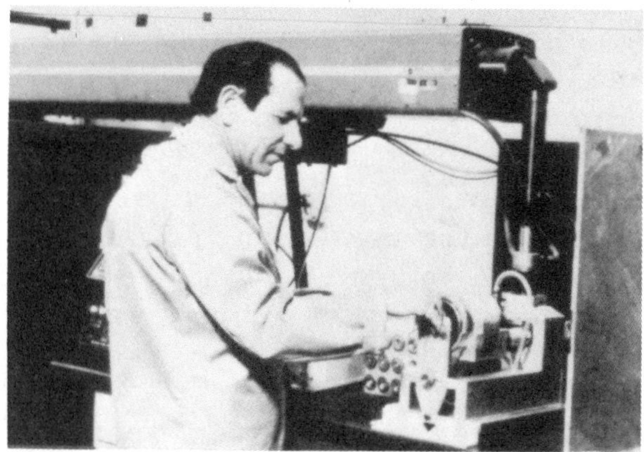

Figure 7. Production Model of Laser Wire Stripper

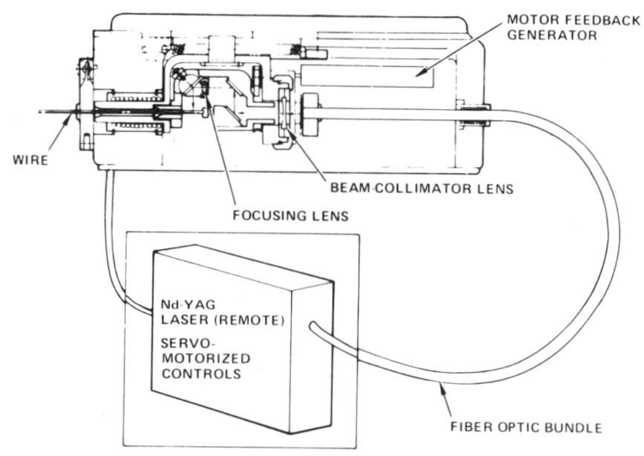

Figure 8. Hand-Held Laser Wire Stripper

The production model of the hand-held Nd-YAG laser wire stripper with fiber optic assembly is shown in Figure 10. The entire assembly is mounted on a portable cart: cooling unit, laser power control, laser, electronic speed control, and counterbalance. With the counterbalance, very little operator effort is required to position the 8-pound stripper.

Oxygen is used for the gas-jet assist, and a vacuum filter system removes debris and vapors. This design also incorporates Rockwell safety requirements.

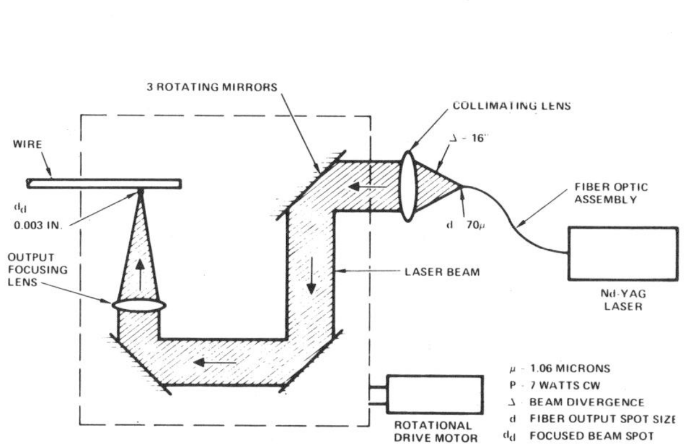

Figure 9. Hand-Held Nd-YAG Laser Wire-Stripping System

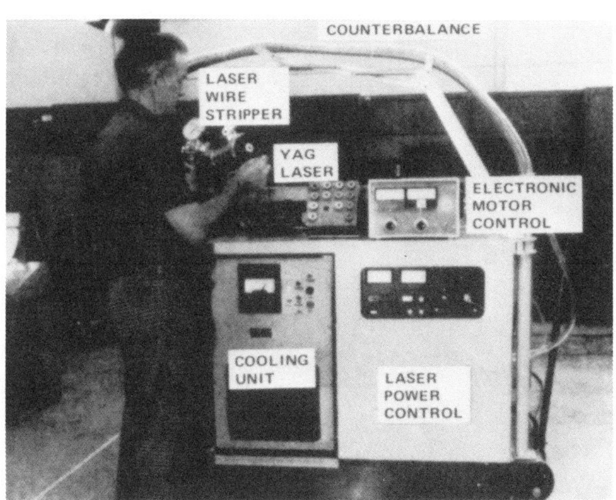

Figure 10. Hand-Held Nd-YAG Laser Wire Stripper With Fiber Optic Assembly

CONCLUSIONS

The focused laser beam is an efficient, reliable means of stripping insulation from wire. This high-speed process requires little setup time and is effective in removing insulation such as PVC as well as Kapton. The process has been proven on the Space Shuttle Program, where it has been used to strip Kapton-insulated wire, the baseline wire in the Shuttle orbiter.

The success of this process is apparently due to the difference in the absorption characteristics of insulation and copper wire when these two materials are subjected to the energy of a laser beam. Since insulation absorbs much more of this energy than does copper, it melts or vaporizes before the copper conductors are damaged. The focused-laser process also offers another advantage: it avoids direct mechanical contact between the wire and other agents.

CO_2 and Nd-YAG lasers are exceptionally reliable tools that can be used in production with repeatable results. Rotational laser stripping is accomplished very rapidly, and longitudinal slitting of the insulation permits the manual removal of the slug; no mechanical puller is required. All metallurgical studies indicate that the metal wire is not adversely affected by the laser beam during stripping. And the strippers can be operated easily by a technician who has received only eight hours of training in the laser wire-stripping process.

Product Marking With ND; YAG And CO_2 Lasers

By M. J. Weiner
Manager, Material Processing Systems
Korad Division of Hadron Inc.

Product marking is frequently necessary for identification, product information, or theft prevention. The unique characteristics of laser marking are discussed -- engraving, dot matrix marking, and mask imaging. Types of laser equipment, cost considerations, and applications for laser marking are discussed in the paper. This paper was originally presented to The Society of Photo-Optical Instrumentation Engineering in San Diego, California, August 24, 1976 (86-05).

Introduction

Historically, the need to mark products has been amply demonstrated. The earliest stone carver marked his wares with a trademark, and today, chip capacitors as small as one millimeter square are marked with product information as well as with the manufacturer's trademark. Lasers have only recently been applied to product marking. When the items to be marked are small, fragile, or must be marked at great speed, laser marking is frequently successful.

Often, marking is desirable for identification. It may be desirable to mark the name of the manufacturer, the batch number being processed, or the serial number of each individual item. Items may be marked to indicate product information in order to communicate how the product is to be used, assembled, or when the object should be removed from the sales shelf. Items can be marked to promote theft prevention. Burglars have found that owner-identified objects are much more difficult to dispose of, and thus are burgled less frequently. Also, recovered stolen goods are more easily returned to the rightful owner.

Lasers have unique characteristics that lend themselves to the marking and serializing of items:

1. The ability to micromachine -- as small as 12 microns in width.
2. The ability to be imaged by using masks.
3. The ability to be manipulated at high speeds (for high speed "writing").

4. The ability to be easily automated.

Laser marking systems have a relatively high cost when compared to conventional marking equipment. For this reason, we will see that although lasers can mark almost anything, the items to be marked should have certain specific characteristics in order to result in an economical marking process.

Methods of Laser Marking

Before discussing lasers for marking, it is useful to identify the three common techniques of laser marking -- engraving, dot matrix marking, and mask imaging.

Laser engraving is similar to mechanical engraving in that the laser is used to micro-drill a groove in the workpiece. Typically, the laser vaporizes the workpiece material and leaves a suitable trench, which may be from 12 microns to 1 millimeter wide, and is typically as deep as it is wide. Laser beam manipulators are capable of moving the beam so as to print alpha-numeric characters in any configuration.

Dot matrix marking is accomplished by vaporizing a series of minute holes (typically 75 microns in diameter) at rates of up to 4,000 holes per second. Rapidly moving mirrors deflect the laser beam and cause the holes to trace out alpha-numeric characters at a rate of 30 characters per second.

Figure 1 shows a laser marking system which can be used for dot matrix or engraving marking. Figure 2 shows a piece of anodized aluminum which has been marked with a 9 x 7 dot matrix. Each hole in a character is located in one of the 63 positions in a 9 x 7 rectangular array. This kind of character is suitable for markings which are 3 mm high or less because the holes appear to merge into lines. For larger characters, this system can provide more dots per character.

Mask imaging marking takes advantage of the parallel (collimated) beam which is produced by most lasers. The beam illuminates a mask which acts as an object for a projecting lens. The lens produces an image of the mask on the workpiece, usually reduced in size. If the laser energy is high enough to remove material or otherwise change the workpiece, a permanent image of the mask is produced. When usable, this technique is highly productive in that marking can be done in one laser pulse -- sometimes taking only a few microseconds.

Types of Lasers for Marking

Both Neodymium:YAG and CO_2 lasers have been used advantageously for laser marking. We should note, however, that any BRH Class IV laser, such as a pulsed ruby laser, can be used for marking under limited conditions.

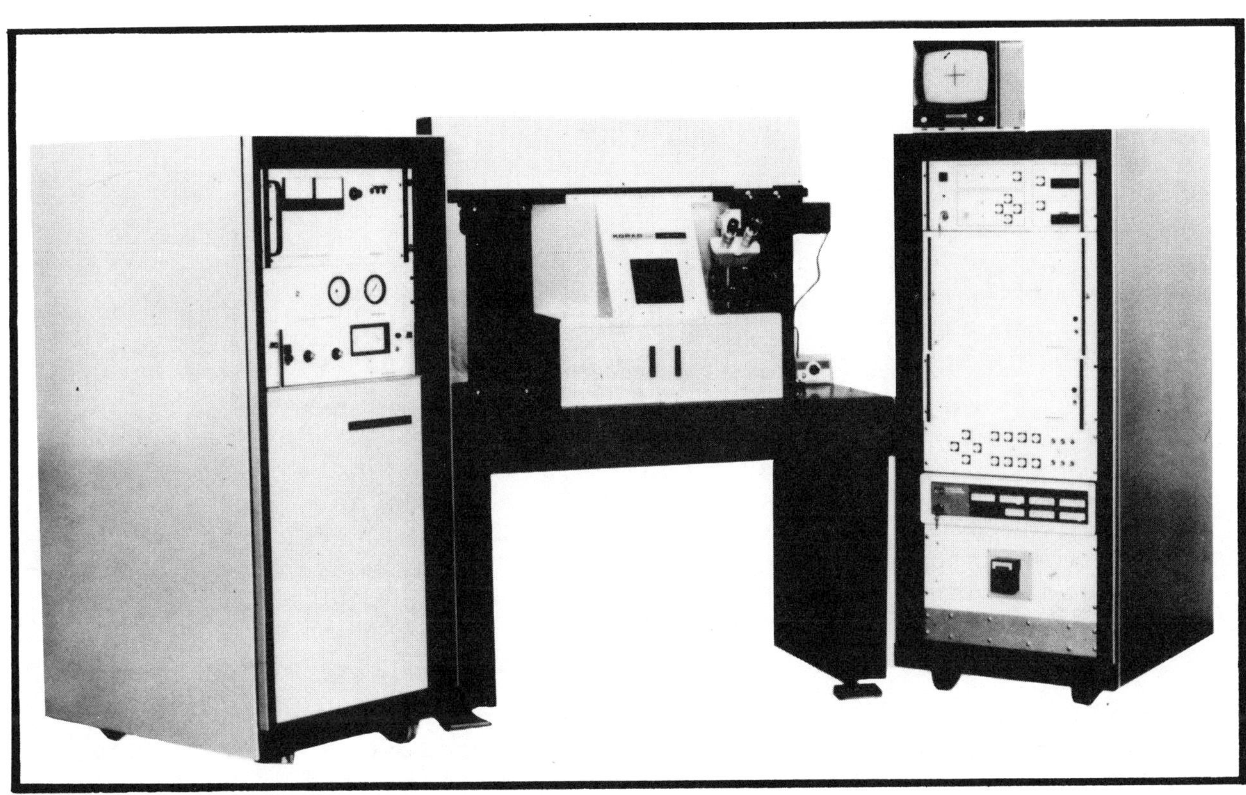

Figure 1 - A laser micromachining system which can be used to laser-mark areas as large as 12 centimeters square. The laser power supply and cooler are shown in the left console, the workstation with binocular microscope or TV viewing and its interlocked access doors shown in the center, and the minicomputer and electronics and controls are in the right hand console.

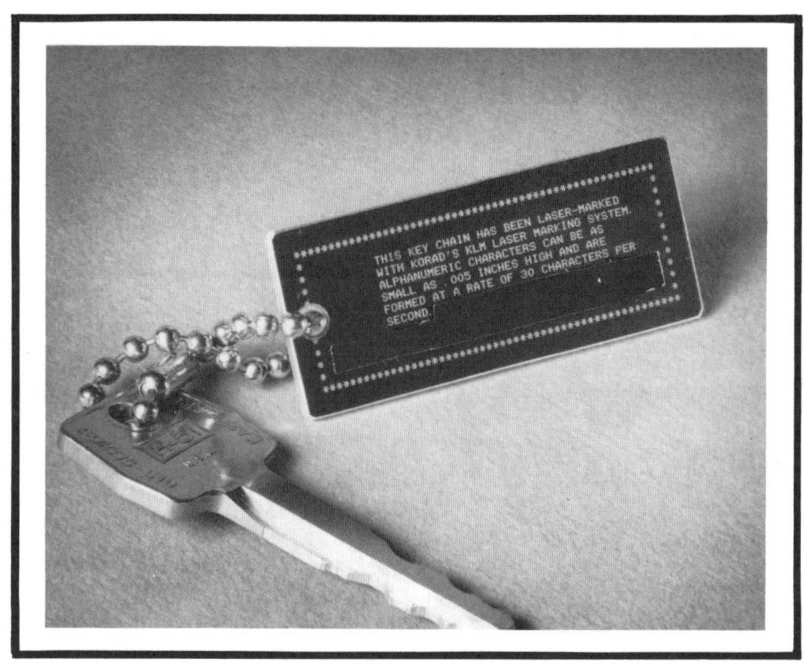

Figure 2 - *An anodized aluminum alloy plate marked by a Q-switched YAG laser using the dot matrix mode. The marking of non-metals, such as the removal of the anodize (the dyed aluminum oxide film) can also be accomplished by using mask-image marking, typically with the aid of a CO_2 laser.*

The Neodymium:YAG laser can be used for engraving. The laser can be operated either continuous wave or Q-switched, and the vaporizing action can often be enhanced with a coaxial jet of oxygen gas. The oxygen enters into an exothermic reaction with certain metals, including the iron based and titanium based alloys.

Dot matrix marking is usually done with a Q-switched Neodymium:YAG laser. The acousto-optic Q-switch allows high peak power, 200 nanosecond pulses to be generated. The system can be commanded by a computer to issue precisely timed pulses at rates to 4,000 pulses per second.

Q-switched YAG lasers can also be used for engraving. The laser is pulsed at a continuous 10 to 30 kHz, and each pulse vaporizes about 1 to 100 nanograms of material, thereby producing an engraved trench.

YAG laser radiation is absorbed by both metals and non-metals to a useful degree, so marking of nearly all materials can be accomplished. However, where non-metals only are to be marked, the CO_2 laser should be considered because of the nearly total absorption of CO_2 energy by most common non-metals.

When operated in the pulsed mode, CO_2 lasers can be used to engrave non-metals. Pulse rates up to 1 kHz are useful, and standard 200 to 500 watt longitudinal flow, longitudinal excitation CO_2 lasers can be used.

Mask image marking usually requires laser pulses with 1 to 10 joules of energy content. The transverse excited atmospheric (TEA) CO_2 laser is capable of generating suitable pulses at repetition rates of up to 10 Hz.

Cost Considerations for Laser Marking

Because laser equipment is relatively expensive -- typically $10,000 to $100,000 for a marking system -- not all applications are economically feasible. Usually, if steel stamp marking will perform satisfactorily in production, laser marking will be difficult to justify. It has been found that an application must have one or more of the following characteristics in order for laser marking to be economically justifiable:

1. Small characters, smaller than achievable by conventional methods, must be required.
2. A permanent and distinctive mark is required.
3. The workpiece cannot be steel-stamped because is is too precise, delicate, hard, or awkwardly shaped.
4. Marking is required without making contact with the workpiece.
5. Marking must be done at exceedingly high speeds.
6. Marking must be done with extreme positional accuracy.
7. Chemical contamination from inks must be avoided.
8. Marking is part of an automated (usually computer-controlled) total process.

Applications for Laser Marking

Semiconductor wafers (silicon, gallium arsenide, etc.) have been engraved/marked with a Q-switched YAG laser in order to identify the manufacturer, batch number, material, diameter, thickness, dopant, crystallographic structure, and to provide other useful information. The markings are made as a series of binary-coded dashes which are machine-readable. In this way, not only the writing, but the reading of the information is completely automatic. In addition, alpha-numeric information can be written on the wafers so as to aid visual identification, as shown in Figure 3.

Serial numbers can be engraved on hand guns with a Q-switched laser. It is possible to mark the gun after it has been assembled when it would not normally be able to withstand

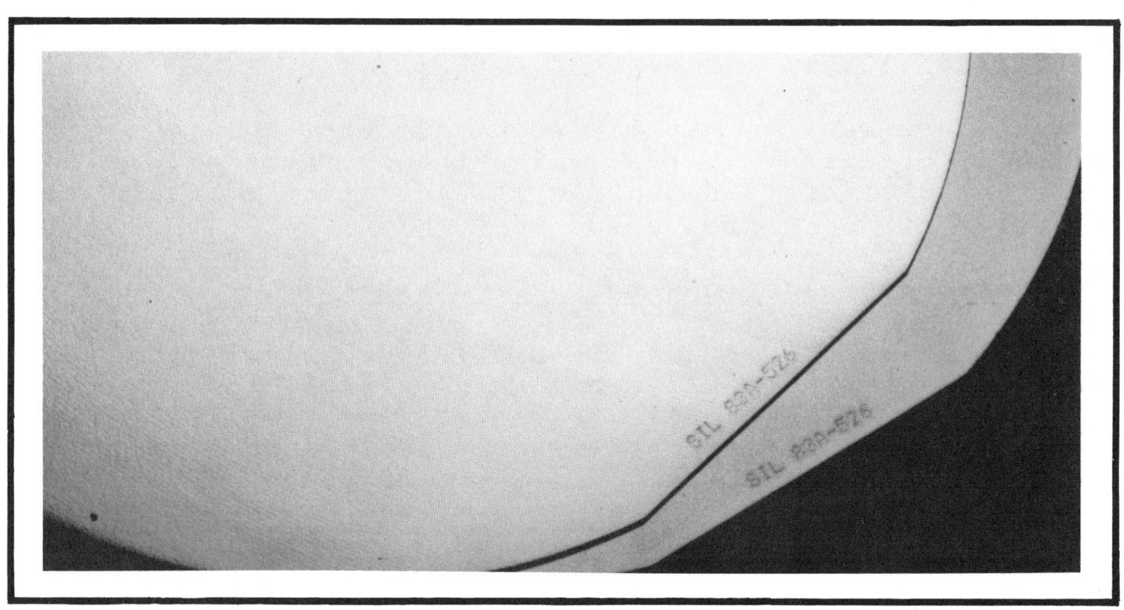

Figure 3 - *Portion of two silicon wafers like those used by the semiconductor industry in the making of integrated circuits. The wafer shown is 75 millimeters in diameter and .037 millimeters thick. The marking is positioned adjacent to the flat of the wafer.*

the stresses of steel stamping. A laser makes a permanent mark which cannot be easily duplicated.

Gold and silver jewelry and precious stones, including diamonds, have been marked by a Q-switched YAG laser operating in both the engraving and dot matrix modes. Figure 4 shows a 14-karat gold-filled locket and a 14-karat gold button which has been laser-marked in the engraving mode. Decorative markings of this type may be more economically accomplished by conventional engraving equipment in the hands of a skilled craftsman. However, the laser is capable of marking valuable items with identification information in nearly invisible characters as small as 125 micrometers in height. Such markings may be highly desirable for theft prevention/owner identification purposes. In the case of precious stones, the owner's driver's license number can be marked on the girdle of the stone in order to permanently identify the gem. Only recutting of the gem could eliminate the mark; and this, of course, would greatly depreciate the value of the gem.

Small chip capacitors, perhaps 2 mm square, have been marked with the manufacturer's symbol and the capacitance

Figure 4 - 14 karat gold-filled locket and 14 karat gold button laser-marked in the engraving mode.

value by using a Q-switched YAG laser. Larger capacitors can be marked with a TEA CO_2 laser in the mask-image mode.

Fully assembled typewriter frames are being marked with a continuous wave YAG laser in the engraving mode. The beam is scanned over the typewriter frame to produce a seven segment display. Once the frame has been assembled, it is too fragile to be steel-stamped.

Railroad car wheels can be marked with serial numbers and other identifying information by using a Neodymium:YAG laser with oxygen assist. Large characters, typically 20 mm high, can be marked on the wheels without structural damage. If wheels with a particular sequence of numbers should develop a crack, the entire batch of wheels with those numbers can be easily inspected.

Consumer packages can be marked with the "pull date" -- the date which the perishable package should be removed from the shelf -- as the package moves down the process line. A TEA CO_2 laser can mask-image mark the package at rates to 10 packages per second.

Large sheets of thin gauge steel can be marked with assembly information and job number coding as they are processed in the production equipment of a large sheet metal fabricator. A Neodymium:YAG laser operating in the engraving mode can produce marks approximately 200 microns in width and depth.

Heavy steel plates, still red hot from the rolling process, can be marked with a Neodymium:YAG or CO_2 laser equipped with oxygen assist.

High volume parts can often be profitably inspected by automatic inspection equipment. Each inspection characteristic can be assigned a number, and if a defect is found, that that number can be automatically laser-engraved onto the part in question. Figure 5 shows an aluminum alloy automobile piston which has been laser-marked in the engraving mode. The engraved letters "9802" might indicate that this piston has an oversized oil ring groove and is excessively out-of-round. As a consequence, material review personnel can dispose of the defective pistons with great ease.

Precision tool and die components are frequently made in small volume and require detailed marking of product information which will disturb the dimensions or structural integrity of the parts. Laser marking can be used here with many obvious advantages over the metal stamping process.

Ceramic parts, prior to glazing, can be marked by a Neodymium:YAG or CO_2 laser operating in the engraving mode. Product information and serial numbering can thus be applied to both precision ceramic parts and high volume pieces such as spark plug bodies or catalytic converter parts.

Laser Marking System

Figure 6 depicts a block diagram of a fully automated laser marking system. The Neodymium:YAG laser is capable of generating over 15 watts of TEMoo laser power at 1.06 microns wavelength. The laser head, containing YAG rod and Krypton arc lamps, is serviced by a power supply and water cooler. A

Figure 5 - An aluminum alloy automobile piston marked in the engraving mode.

Q-switch driver delivers RF power to an acousto-optic Q-switch, which allows the laser to generate pulses with peak powers approaching 10^9 watts per square centimeter. These pulses are gated on and off by the Q-switch in accordance with an input from the minicomputer.

The laser beam is directed to a galvanometer beam positioner which sweeps the beam over a 5 centimeter by 1.8 centimeter area, as directed by the minicomputer via the computer interface and beam positioner amplifiers. A standard teletype serves as the input/output for the system. Although other I/O devices can be used, the low cost and ability to obtain a hard copy record of the input are highly advantageous in a system of this type.

The entire system is enclosed in accordance with the Bureau of Radiological Health safety requirements which became effective for all laser systems produced after August 2, 1976.

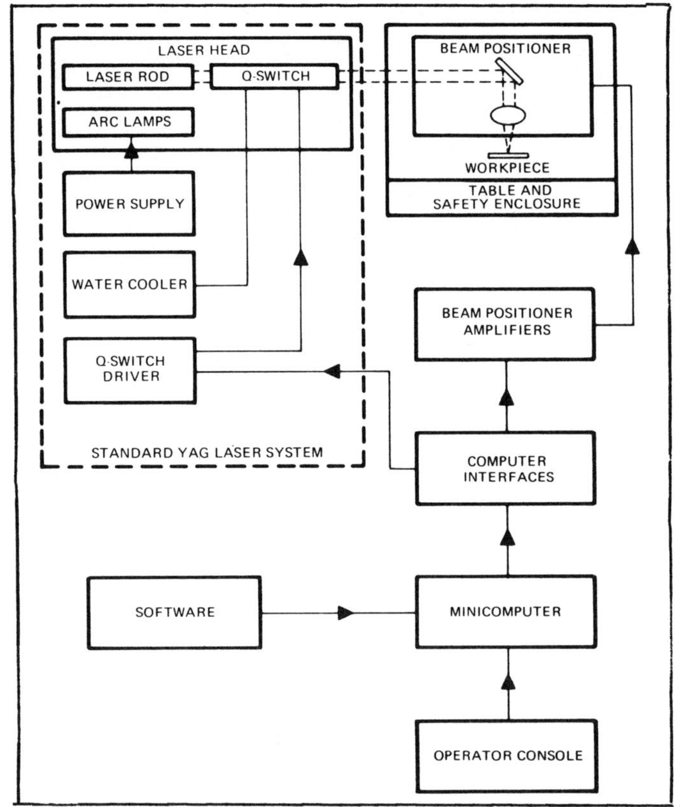

Figure 6 - Block Diagram of a Fully Automated Laser Marking System

Figure 7 shows a photograph of the completed system. The laser power supply, cooler, and Q-switch driver are located in the cabinet to the left. The laser head with its optical train and safety enclosure are on top of the table, and the minicomputer, interface electronics, and beam positioner amplifiers are located in the table base along with an optional high speed paper tape reader. The standard ASR-33 Teletype is not shown in the photograph.

Summary and Conclusions

Product marking is often required for identification, product information, or theft prevention. Engraving, dot matrix, and mask imaging are three useful techniques of laser marking. Neodymium:YAG lasers are suitable for both metal and non-metal marking; CO_2 lasers are most useful for non-metals marking.

There is a wide variety of suitable laser marking applications, but the relative high cost of the marking equipment eliminates those applications which can be readily handled by conventional techniques. Nevertheless, it is the opinion of the author that enough suitable applications for laser marking exist so that well in excess of 100 laser marking units will be in production before 1980.

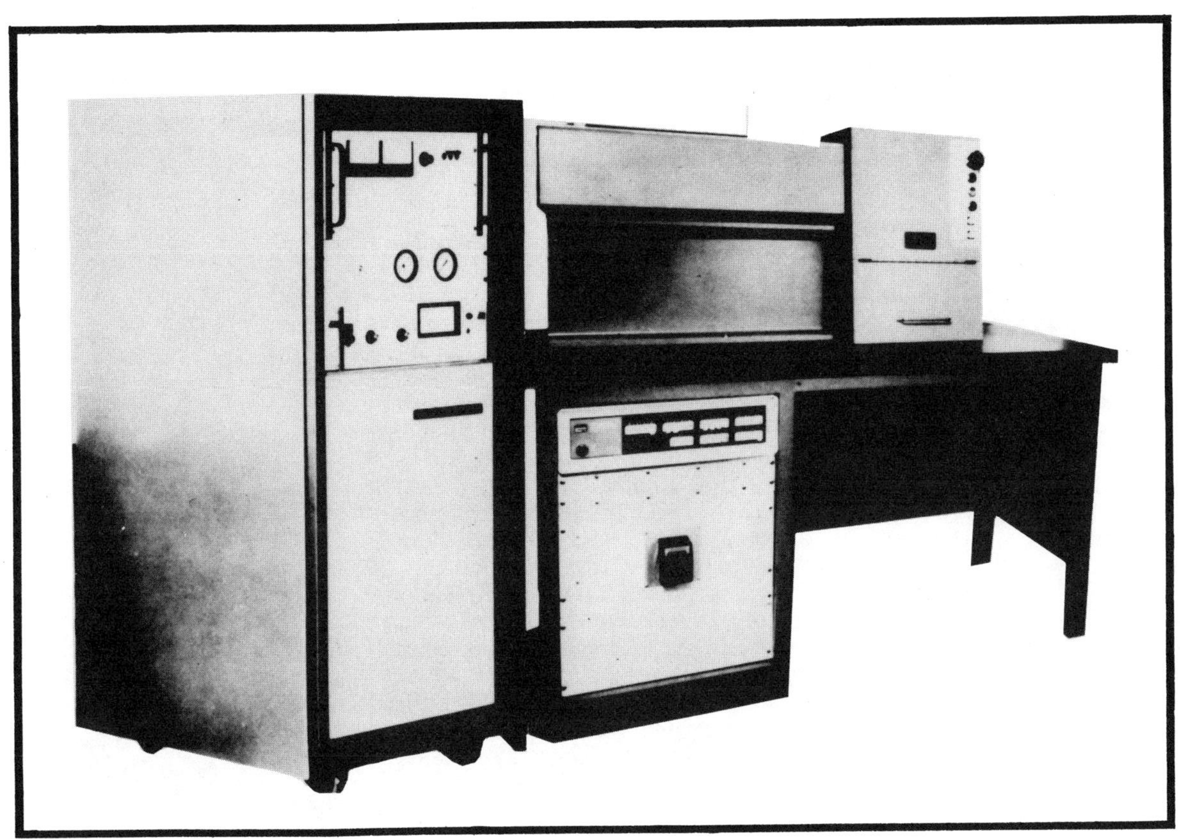

Figure 7 - The KLM Laser marking system. Laser power supply, cooler, and Q-switch driver are housed in the console on left. The laser and safety enclosure are on top of the table, and minicomputer, beam positioner, and electronics in table base.

Reprinted from: Metal Progress, May 1975

Laser Processing at Ford

By MICHAEL YESSIK and DUANE J. SCHMATZ

Automotive welding, cutting, drilling, and hardening applications are just down the road. Powerful CO_2 units will be tapped for their ability to handle larger parts at the industry's high production rates.

POTENTIAL automotive laser processing applications fall into three categories: welding, cutting and drilling, and surface hardening. Because the mechanisms for these processes are basically thermal in origin, the laser can be considered as simply a heat source.

The main distinction between the laser and most other heat sources is its ability to obtain very high power densities, on the order of 10^7 W/in.^{2} ($\approx 10^6$ W/cm^2) or greater.

Another advantage is the ease with which beam (heat source) size and position can be controlled. The beam is focused with a lens, or mirror, and the focused spot size is readily controlled by the choice of lens and the distance from it to the workpiece. The spot can be made as small as a few thousandths of an inch ($\approx$0.1 mm). Beam direction can be controlled with small, lightweight mirrors. This makes it easy to automate laser processes — the beam can be made to move over programmed paths of any complexity.

The electron beam can rival the laser in power density, but not in beam controllability.

Another advantage over an electron beam is that the laser beam is not attenuated in air and can be sent over very long distances from the laser to the work area. This means that more complex paths can be followed and opens up the possibility of time sharing a central laser system with several work stations.

Finally, the laser produces no penetrating radiation. Protective shielding is very simple and, in fact, is usually made of a plastic such as Plexiglas.

CO_2 Lasers — To qualify as an industrial processing tool, a laser must also be able to maintain a relatively high average power output in addition to producing a high power density.

Continuous carbon dioxide lasers top the power output list and are also generally simpler and less expensive. Admittedly, other laser types, both pulsed and continuous, have been used by industry. However, the applications have been for parts much smaller, or at production rates much lower, than those of interest to automakers.

This discussion is consequently limited to the application of continuous CO_2 lasers with maximum power levels of roughly 100 W to 20 kW. All are commercially built units designed for industrial processing, and all operate on the same basic principle: a mixture of CO_2, N_2, and He gases flowing through an optical resonant cavity is excited by an electrical discharge. The output beam has a wavelength in the infrared region (10.6 μm) and is not visible.

Welding

In its simplest form, laser welding occurs when enough power is applied to melt the metal surface. The molten zone takes on a radial configuration as in conventional arc welding.

However, when the power density rises above a certain threshold level, keyholeing — a deep penetration effect — occurs. Here, the material at the most intense part of the beam vaporizes and the adjacent regions melt. As the high vapor pressures generated push molten metal to the side, out of the beam path, the beam continues to deepen the hole.

The vaporization is not intense enough to blow molten metal out of the hole. And an inert gas cover is provided to prevent oxidation reactions. These influences cause a balance to be set up between the pressure created in the hole and the head of molten metal being supported. This determines

Fig. 1 — Deep penetration is a feature of laser welding. These 0.032 in. (0.8 mm) low-carbon steel sheets were joined at a laser power of 5.2 kW and a weld speed of 240 in./min (102 mm/s). Total penetration: 0.105 in. (2.7 mm).

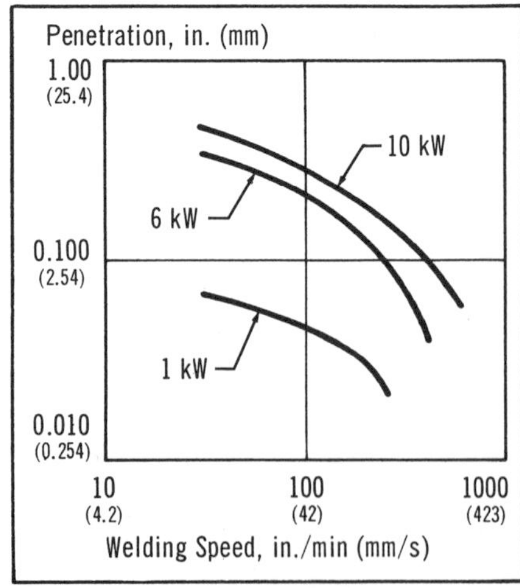

Fig. 2 — Weld penetration depends strongly on both laser power and weld speed. These units were manufactured by Photon Sources Inc. (1 kW), United Aircraft Research Laboratory (6 kW), and Avco-Everett Research Laboratory (10 kW).

Messrs. Yessik and Schmatz are members of the Scientific Research Staff, Ford Motor Co., Dearborn, Mich.

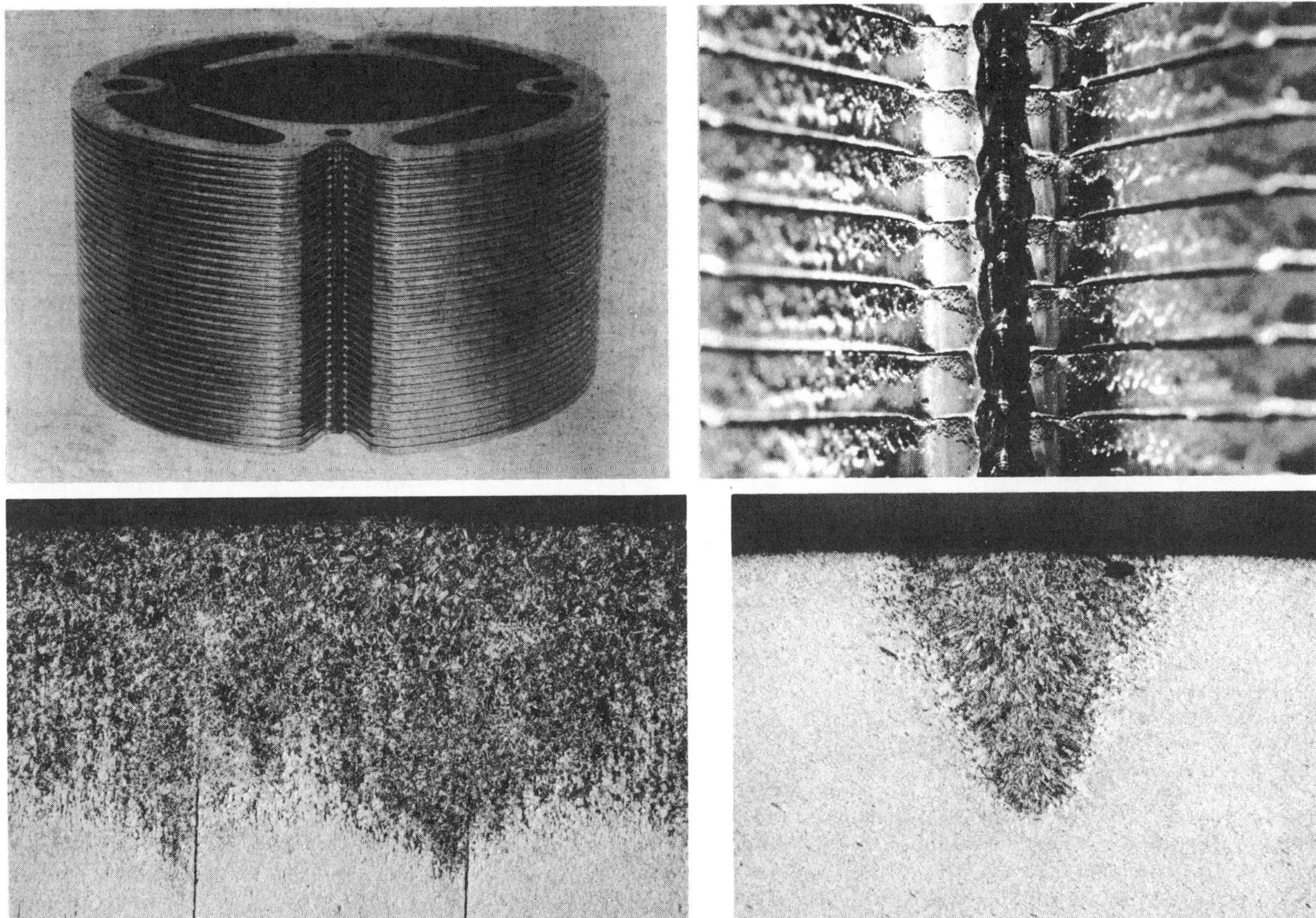

Fig. 3 — The heater motor's yoke assembly (a lamination stack) is a potential laser welding application. Shown are (clockwise from top left) a welded yoke, a top view of the weld bead, and 25× transverse and longitudinal weld sections. L. Mihalyfy of Ford's General Products Div. is credited with the idea.

depth of penetration.

As the beam or workpiece moves, molten metal flows in behind the hole and cools to make the weld. The result: very deep and narrow welds.

The deep penetration threshold region lies between 500 and 1000 W. Most of our welding has been done in the 1000 to 10 000 W region. A cross section of a typical high power laser weld is shown in Fig. 1. Note the high depth-to-width ratio. Welds of this type can be made only with electron beams and lasers — no other method produces sufficiently high power densities.

In addition to producing deep welds, the keyhole effect also minimizes the problem of beam reflection from shiny metal surfaces. Once it's created, the hole behaves almost like a black body and absorbs essentially all of the beam.

Plasma — A problem can occur if the evolved metal vapor initiates gas breakdown and plasma generation in the region of high beam intensity just above the metal surface. The plasma — highly absorbent to laser energy — can block the beam and inhibit melting. To prevent plasma buildup, an inert gas jet is often directed along the metal surface. Properly designed jets effectively remove the plume, and, as mentioned above, protect the surface from oxidation.

Laser welding boasts advantages common to other laser processes: no contact required with the work, small heat-affected zone, easy beam manipulation, and high processing speeds. In addition, its deep penetration capability eliminates the need for a filler metal.

Cooling Rate — Penetration is a function of welding speed, beam shape and focusability, and the properties of the material being welded. Typical penetrations in steel for various welding speeds are plotted in Fig. 2 for three different lasers. A weld more than 0.5 in. (12.7 mm) deep can be made at 10 kW; while, at 1 kW, welds 0.08 in. (2.0 mm) deep or deeper can be made.

Weld quality not only hinges on sufficient penetration, but also on the effects the laser beam has on the base material's structure. In general, heating and cooling rates are much higher in laser than in conventional welding, and heat-affected zones are much smaller. These rapid cooling rates can, for example, cause cracking and increased notch sensitivity in carbon steels as their carbon content increases. For these reasons, laser welds are generally restricted to low-carbon grades.

On the other hand, a rapid cooling rate is advantageous when welding high-strength, low-alloy steels. By judiciously selecting

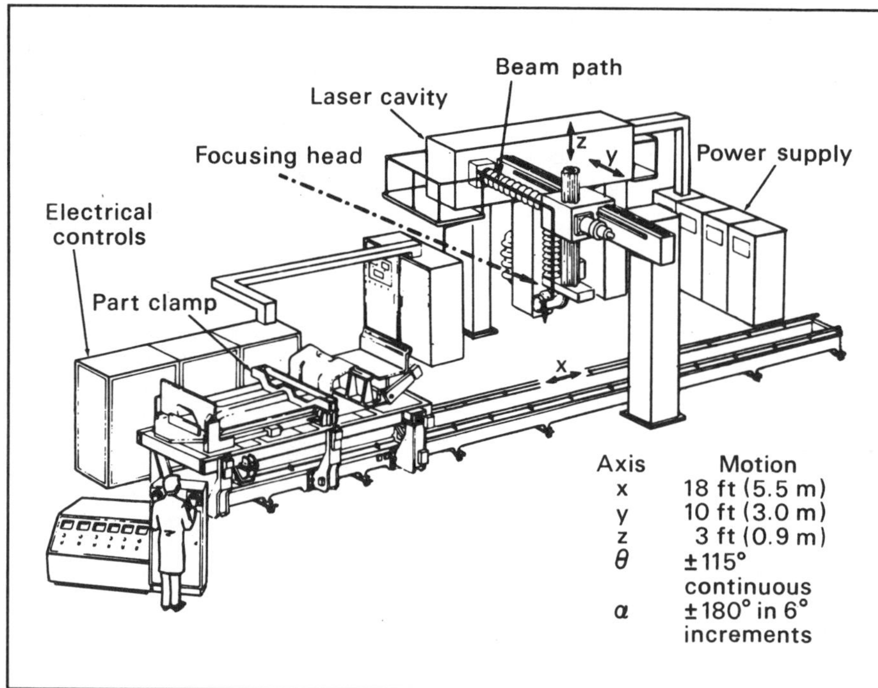

Fig. 4 — Hamilton-Standard supplied this schematic of Ford's laser underbody welding system. The development of this system and the automatic laser hole cutting system (Fig. 6) is largely due to the efforts of E. E. Lohmolder of Ford's Automotive Assembly Div.

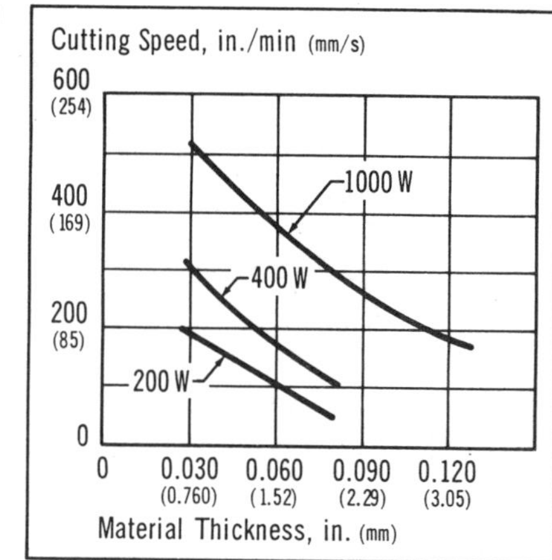

Fig. 5 — The increase in cutting rate with laser power is roughly linear. These data are for oxygen-assisted cutting of low-carbon steel.

welding parameters, laser welds can be made in HSLA steels that have properties essentially identical to those of the base metal. In conventional arc welding, a higher total heat input and a slower cooling rate can cause a strength decrease.

There is a drawback: the laser weld cools so rapidly at the highest welding speeds that bead hardness approaches the maximum obtainable for the particular steel's carbon content. As a result, it's also preferable to select lower carbon content HSLA grades.

Aluminum is another material that lends itself to laser welding. The localized heating helps keep mating surfaces clean and minimizes the buildup of oxide which tends to prevent wetting.

Motor Yoke — A potential welding application is shown in Fig. 3. This heater motor yoke assembly is made up of a stack of laminations joined by a 1 kW laser. Stack height is 1.5 in. (38 mm) and the thickness of each lamination is 0.050 in. (1.3 mm). This stack was welded at a speed of 70 in./min (300 mm/s) so that two weld passes could be made in a total time of 3 s, well within the allowed time for this application.

The weld bead (Fig. 3, top right) has a very smooth appearance with no apparent porosity. Longitudinal and transverse sections through the weld (Fig. 3, bottom) reveal a weld depth of about 0.050 in. (1.3 mm), the same as the lamination thickness, and negligible porosity and no cracks. The configuration could be welded by either moving the beam or the part. The laminations do not require cleaning because a small amount of oil or other contaminant does not adversely affect weld quality.

Underbody Panels — An example of a higher power application is the 6 kW underbody welding system (Fig. 4) recently built for Ford by Hamilton-Standard. In this system, stamped panels are loaded in an assembly fixture which is part of a transfer table. The table provides one axis of weld motion (the x-axis in Fig. 4). The platform-mounted laser has its beam directed through a transport system which provides four axes of mo-

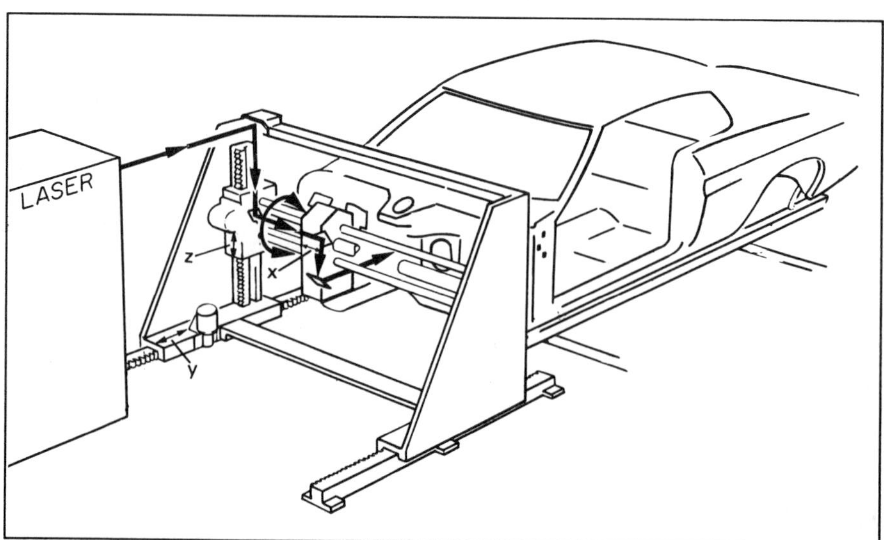

Fig. 6 — Improved precision is a major advantage of this setup for drilling dash panel option holes. The laser beam can be translated along the x, y, and z directions; and it can be rotated about the x axis.

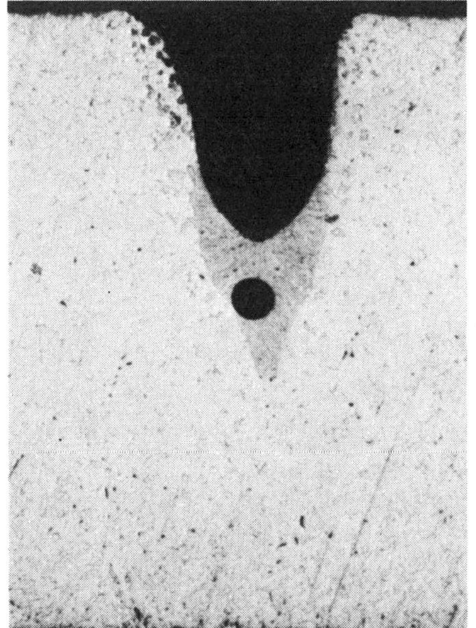
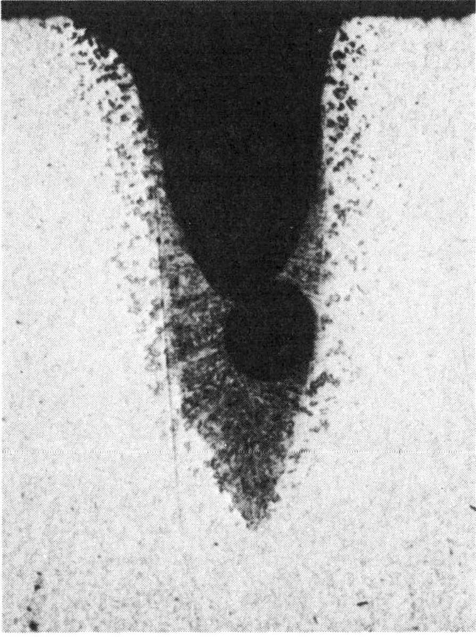
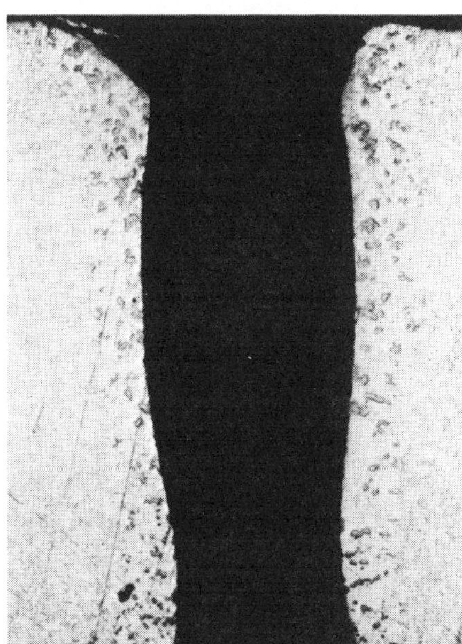

Fig. 7 — Oxygen-assisted laser pulse times of 20 and 50 ms (left and center) do not give complete penetration in 0.074 in. (1.9 mm) thick low-carbon steel. A pulse time of 110 ms (right) does.

tion. Two are translation, y and z-motions; and two are rotations, one in the horizontal and one in the vertical plane.

Beam motions (performed by moving mirrors), transfer table motions, and clamping and unclamping systems are directed by a computer which allows welding along preprogrammed paths at selected weld speeds.

Most underbody welds are of the burnthrough type joining two 0.035 in. (0.9 mm) thick sheets at a speed of 400 to 450 in./min (170 to 190 mm/s). The weld configuration is similar to that shown in Fig. 1 where four sheets are joined by a burnthrough weld.

The laser underbody weld is continuous, as opposed to current spot welding practices, and results in high structural integrity and eliminates the need for a sealing operation. The system's programmability means that underbodies for any car line or model can be welded simply by calling up the correct program from the computer's memory.

The system is currently being evaluated for the welding of prototype 1975 Torino and Montego underbodies.

Cutting and Drilling

Laser cutting is effective on a wide variety of materials including metals, ceramics, plastics, vinyls, rubbers, and composites. Fixturing is simplified because there's no physical contact with the part; cutting rates are generally quite fast; and there's no limitation on cut configuration.

Inside cuts, of any shape, can also be readily made. And, because the beam can be focused down to a very small spot, kerf widths can usually be made as narrow as a few thousandths of an inch ($\approx$0.1 mm). Cut edge surfaces are generally very smooth and clean thanks to the laser's highly localized heating.

Cutting mechanism depends somewhat on the material being cut. For nonmetals, the beam is highly absorbed and its high power density results in an almost explosive vaporization at the region of maximum intensity. This causes the surrounding molten material (and in some instances particulate matter) to be blown out, greatly increasing the cutting rate over that which would be obtained if vaporization were the only removal mechanism. Cut edge surfaces still remain smooth. As in welding, an inert gas jet is used to prevent oxidation and to help blow away debris.

Example — A glass-filled polyester, 0.130 in. (3.3 mm) thick, is being considered for use as a body panel material. A 400 W laser gives high quality cuts at a rate 60 in./min (25 mm/s). This material is very easily cracked by conventional cutting or drilling.

Another example is silicon ni-

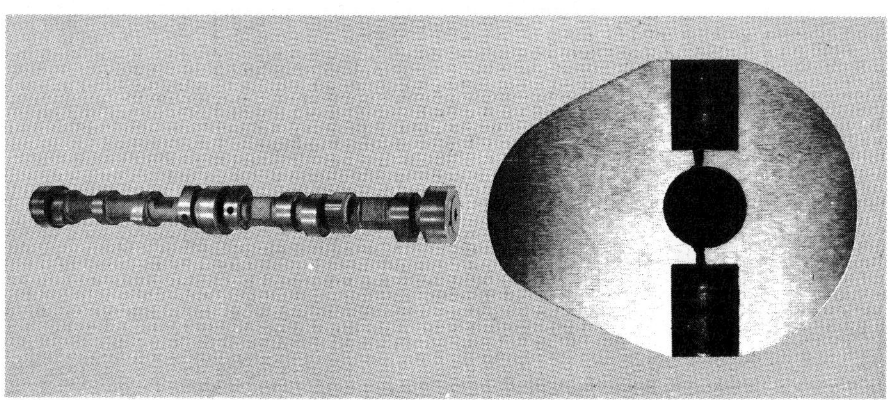

Fig. 8 — The laser makes easy work of drilling oil metering holes in hardenable iron, rifle bore camshafts (left). The two holes (see lobe cross sections) are located between the large central bore and peripheral lead holes.

tride, a very hard ceramic under evaluation for automotive turbine engine applications. A laser can cut it ten times faster than a diamond wheel can. It can also cut any shape — the diamond wheel's limited to straight line cuts.

Cutting of steels and other reactive materials is done with an oxygen gas jet assist located coaxial with the laser beam. The beam provides sufficient heat to initiate and sustain the highly exothermic oxidation reaction. Cutting rates are very high and substantially less power is needed than in laser welding. Oxide products are rapidly swept away by the gas jet and cut surfaces are generally clean. Narrow kerf widths and small heat-affected zones are also characteristic.

Two particularly important advantages of laser cutting of steel are that cutting rate is essentially independent of steel hardness, and there is no tool wear.

Data for low-carbon steel are plotted in Fig. 5 where cutting rate is shown as a function of thickness for various laser power levels. Note that the increase in cutting rate with laser power is roughly linear over the range of 200 to 1000 W.

Dash Panel — The hole cutting system shown schematically in Fig. 6 is an example of a potential production application for laser cutting of steel. The system is designed to automatically cut holes of varying size at various locations in the dash panel. The hole layout is determined by how many of several regular production options have been specified for that unit.

Information concerning the size and position of each hole is read from punched cards. The computer control system then automatically moves the body into the work station, registers the dash panel to the laser cutter, cuts the required hole pattern, and moves the body out of the work station. Beam motion is controlled by four moving mirrors.

System advantages include much greater precision in location and size of option holes, and the ability to cut holes of other-than-circular shape. Sufficiently high cutting speeds could be furnished by a 400 W laser.

Drilling — Laser hole drilling differs from hole cutting in that there is no beam motion involved.

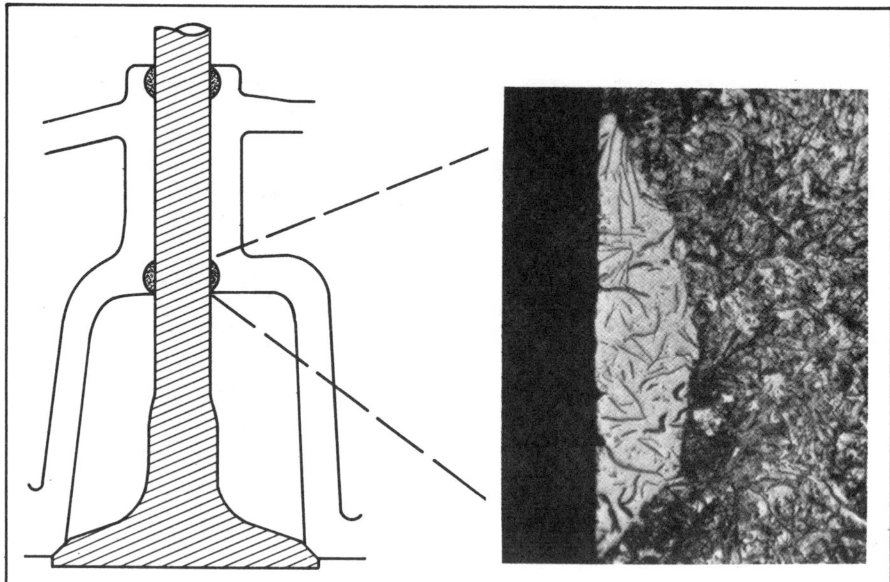

Fig. 9 — The gray iron valve guide schematic shows the locations of two laser-hardened bands. The 80× photomicrograph (reproduced here at 57½%) reveals a hardened area structure of flake graphite in a fine martensitic matrix.

An oxygen jet is used for steel applications to sweep molten oxides, along with some molten iron, out of the hole and leave clean, smooth sides.

Various hole drilling stages are illustrated in Fig. 7. Incomplete penetration has occurred at 20 and 50 ms. Most of the molten material has been blown out of the top of the hole. That remaining on the sides of the hole when the beam was shut off has flowed to the bottom and solidified. Complete penetration is evident at 110 ms. Essentially all of the molten material has been blown out. Hole sides are clean and there's a very small heat-affected zone. Note the hole's shape. Its sides tend to taper in from the top with a slight flareout at the bottom.

Hole size can be varied by controlling beam spot size and the length of time that the beam is on. Tolerances obtainable depend on hole diameter and depth, and on the amount of laser power available.

Camshaft — A potential laser drilling application (Fig. 8) is the oil metering hole in a hardenable iron, rifle bore camshaft. The problem is one of drilling a small hole at the bottom of a large hole. Conventional drilling or punching would be very difficult.

The cross sectional view (Fig. 8, right) shows two holes, one drilled from each side. Each hole was completed in 0.1 s using 375 W of laser power. Hole diameter is 0.030 in. (0.76 mm) and hole depth is 0.065 in. (1.65 mm). Tolerance on the diameter is ±0.005 in. (±0.12 mm). A higher powered laser could be used if an improved tolerance were needed.

Surface Hardening

Laser hardening is basically quite simple because it only requires heating of the surface by some beam absorption.

Ford's interest lies primarily in irons and steels. Beam absorption in these metals is strongly dependent on surface finish and can be enhanced by roughening the surface or by coating it with a thin layer of a highly absorbing material. Depth and width of the hardened zone depend on beam focusing and shape, the rate at which the beam moves over the surface, and the material's microstructure.

Mechanisms — Hardening can occur via either a solid-state transformation or a melting and solidification mechanism, depending on material.

In iron and steel, hardening is a function of the availability of carbon in the microstructure. A fine carbide dispersion in a ferritic matrix, as in pearlitic gray iron, allows hardening to occur in the solid state. In fact, melting would be undesirable here for its embrit-

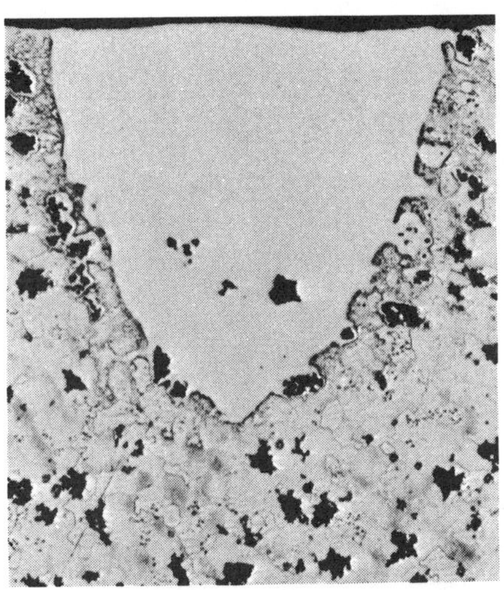

Fig. 10 — The laser can also harden by a localized melting and solidification mechanism. The irregular interface in this hardened band is caused by preferential melting around the ferritic malleable iron's temper graphite. 80×.

tling potential. However, with irons having very little dispersed carbide, such as ferritic malleables, solid-state diffusion from the large graphite particles is too slow to produce the desired solution of carbon in the allotted time. In these cases, melting is essential.

Laser hardening has two major advantages. The first is its ability to selectively harden. This minimizes the total heat input to the material and eliminates the distortion often associated with other hardening techniques. Second is its ability to heat areas, such as small bores or complicated shapes, that aren't accessible to induction hardening. Both advantages stem from the beam's small size and ease of manipulation.

Gray Iron — One application under investigation at Ford is the hardening (for wear resistance) of the inner bores of gray iron valve guides (Fig. 9). Valve guide wear causes increased oil consumption and exhaust emissions, and is particularly severe with the use of lead-free gasoline.

The two hardened rings in this example were made by first positioning a copper rod with a 45°, polished surface in the guide to act as a mirror. The beam directed into the bore is reflected by the copper mirror onto the guide surface. The beam is made to move in a circle around the inside of the guide by rotating the copper rod.

A 400 W beam was spread out (defocused) to a width of 0.060 in. (1.5 mm) and its travel speed was set to give the desired depth of hardening with no surface melting. In this case, hardened depth is 0.010 in. (0.25 mm) and the beam's linear speed over the surface is 10 in./min (4.2 mm/s).

The micrograph (Fig. 9, right) shows the hardened area and also reveals that there is no change in surface smoothness. The hardened area's structure is primarily a matrix of fine martensite containing flake graphite.

Note that by making two bands — at the positions of maximum stress between valve and guide — heat input is minimized and no measurable distortion is produced.

Malleable — An example of the type of hardened region that can be obtained in malleable iron by surface melting is shown in Fig. 10. A focused 400 W beam, at a rate of 120 in./min (51 mm/s), produced a molten region measuring 0.022 in. (0.56 mm) deep and 0.022 in. (0.56 mm) wide.

Careful control of surface temperature accounts for the very smooth surface, even though melting has occurred. Also note the irregular interface caused by preferential melting around the malleable iron's temper graphite. This is in contrast to the regular interface produced by the solid-state transformation in the gray iron example (Fig. 9).

A similarity between the two is that nearly the same cross-sectional area of hardening was obtained, although the hardening rate was considerably higher in the case where melting occurred. It's apparent that process efficiency is much less when the beam is defocused and power density is low because substantially more power is wasted in heating the whole sample.

Laser Cutting For Aircraft Manufacturing

By John Huber
Grumman Aerospace Corporation

The use of a 250-watt carbon dioxide laser to rough-cut aircraft structural parts has been in production at Grumman Aerospace Corporation since 1971. The equipment modifications and tooling requirements to adapt this process to production for both flat and formed parts are discussed. The impact of higher laser power levels on cost and performance is evaluated. Typical applications to a variety of aircraft parts are shown to illustrate the range of applications.

INTRODUCTION

The term "laser" is an acronym for light amplification by the stimulated emission of radiation. The laser is a device which generates a highly collimated light beam that can be focused to a small spot size. When the focused light source strikes a surface, a portion of the light energy is absorbed and performs work in the form of melting and burning. The process has been usefully adapted to cutting both metallic and nonmetallic materials. The carbon dioxide continuous-wave laser is the most common high-power laser used in industry for cutting. The laser radiation produced from this laser has a wave length of 10.6 microns, which is in the far infra-red region of the spectrum. Since the 10.6-micron wave length has about ninety five precent of its radiation reflected by most metallic surfaces, its application to metal cutting can require thousands of watts of power to provide practical cutting rates. The use of a reactive gas in the cutting process made it possible for metal cutting to be accomplished in an economically competitive way by using smaller commercially available lasers in the hundred-watt range.

REACTIVE GAS JET LASER CUTTING PROCESS

The reactive gas, usually oxygen, is introduced coaxially to the focused laser beam through a nozzle at the workpiece surface (Figure 1). The reactive gas assists the cutting process in the following three ways:

- Oxidation of the metal surface in the presence of the hot focused laser spot provides greater energy absorption of the laser radiation by the metal.
- Heat from the exothermic oxidation reaction provides a considerable amount of energy for the cutting process.
- Clearing of molten metal oxide from the kerf of the cut.

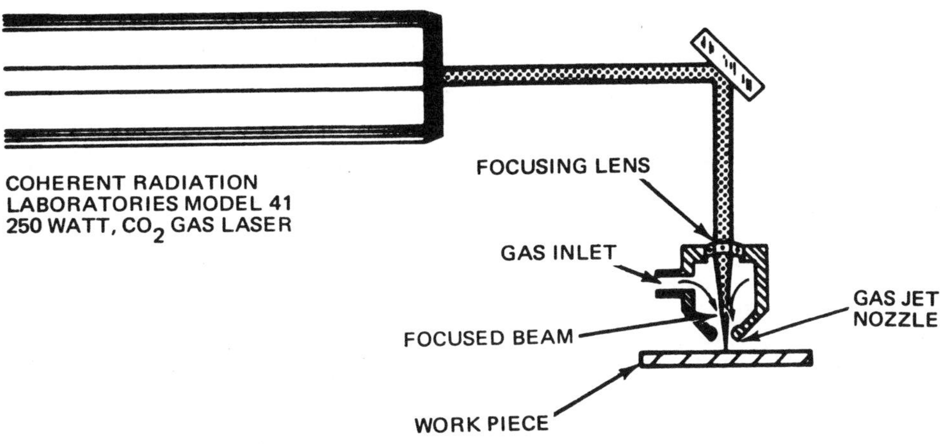

Figure 1

Schematic Representation of Laser Cutting Process

Figure 2

Laser Cutting Facility

The production laser cutting system used at Grumman Aerospace Corporation is shown in Figure 2. The system development was started five years ago to reduce trimming costs associated with titanium. The principal components of this system are:

- Laser - Initiating energy source to start metal burn.
- Optics - Energy focusing mechanism to provide the required density.
- Assist Gas - In the case of metal cutting, it is the reactive gas required to assist the laser by providing the necessary additional energy to make the cut. The gas used is oxygen and the principal parameters are gas pressure and nozzle configuration.
- Positioning - Numerical control system required to direct the laser around the part configuration and to turn laser and reactive gas on and off. Adaptive controlled height sensor is used to provide proper axis motion.

The benefits of such a laser system used as a cutting process are:

- High cutting rate
- Low heat input (minimizes distortion and heat damage)
- No mechanical forces (no mechanical distortion and low holding forces for tooling)
- Minimal waste (small kerf)
- Point source cutting (sharp corners can be cut, not sensitive to directional change and readily adaptable to automated position controls)
- Low operating costs (about 250 watts at $1/hr) for lower power levels (no tools to wear out)

EQUIPMENT MODIFICATION AND TOOLING

Height Sensing - The key to the production application of the laser cutting system is an adaptive height sensing system which provides on-line corrections for variations in tooling, part waviness, part contour and table alignment. Two types of systems were evaluated: non-contact and direct contact (Figure 3).

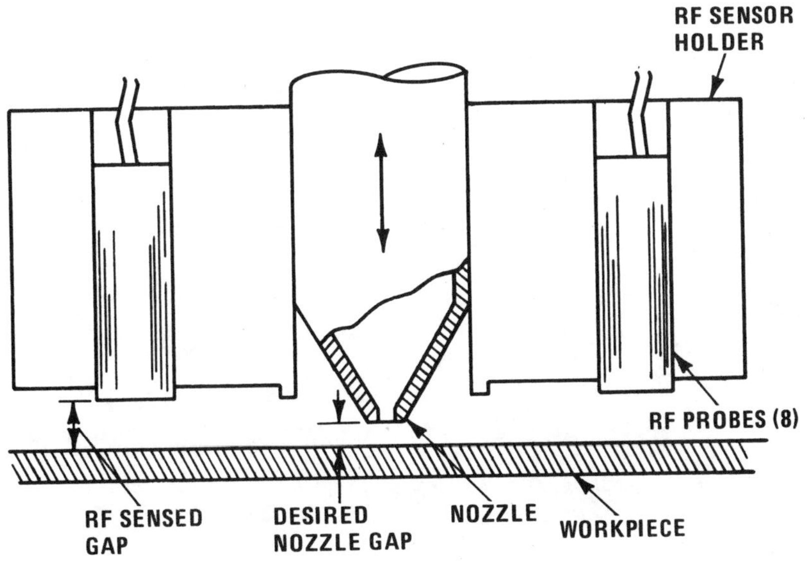

a. NON-CONTACT

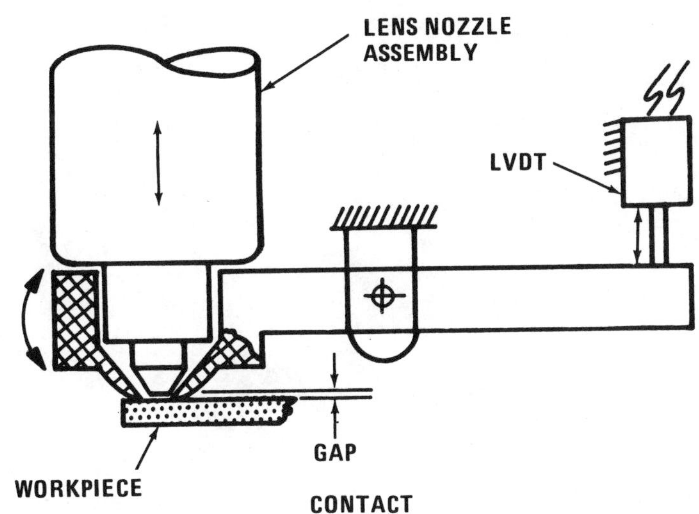

b. ELECTROMECHANICAL

Figure 3

Height Sensor Systems

Figure 4

Manual Rotary Guide System

The non-contact system uses RF-coils operating on an eddy current principal coupled with a closed-loop servo-system to maintain a predetermined distance between the tip of the gas jet nozzle and the workpiece. Coil placement on a 2 1/2-inch-diameter circle around the nozzle introduced errors on cuts made near the edge of a part because some coils were off the cutting surface.

The direct-contact system utilizes a carbide tip, concentric with the gas jet nozzle, floating on the workpiece with a lever arm linked to a linearly variable differential transformer. This sensor is used to measure the difference between the preselected distance from the jet nozzle to the workpiece surface and the actual distance. The signal produced is used to control a closed-loop servo-system that continually seeks to maintain the preset value. The height sensor shown in Figure 3B permits the tracking of part contours whose surface tangents are ± 30 degrees with respect to the machine's horizontal plane. Cutting evaluations with this height sensor have shown it capable of tracking up and down on a 30-degree incline at feedrates of 150 ipm while maintaining a gap tolerance of ±.005 inch over a 10-inch height range.

Manual Trimming - The use of a laser for rough trimming small parts has been demonstrated using a hand-operated guide. The system permits the use of templates to define the cutting path and provides the necessary controls to use a laser in a manner similar to that for a router or a saw. The system shown in Figure 4 operates in the following manner: The motor-driven, rotating guide (Figure 5) provides the operator with a means of manually passing the workpiece beneath the laser head at a preselected feed rate. Path direction is controlled by an overlay template buttoned to the workpiece. The heart of the system is the rotating guidepiece which incorporates a 0.5-inch setback (guide surface to laser center line). The rotating guidepiece provides a smooth and continuous manual feed as the operator pushes the workpiece/template packup against the guide. The guide is water-cooled, driven by a variable-speed motor, and surrounds the existing laser beam to collect and conduct exhaust gases and oxides away from the operator. The operator is not exposed to the products of combustion or the laser beam. One of the principal advantages of this system is that it obviates the need for both a transport carriage and position control system -- two of the major cost items in the laser cutting system.

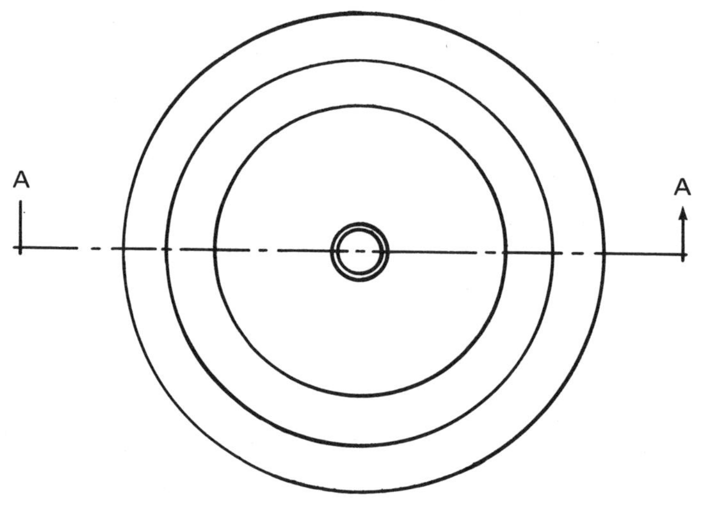

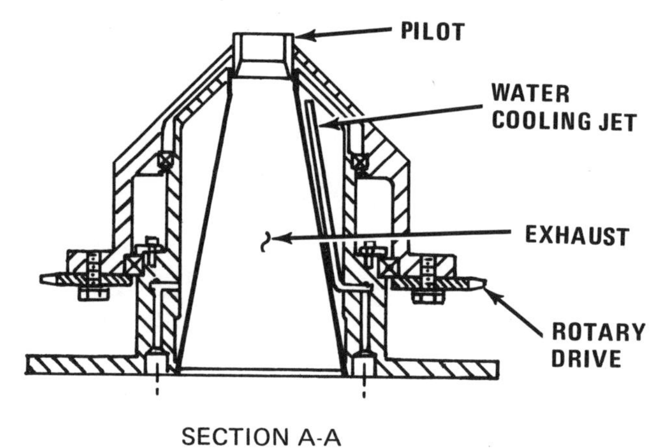

SECTION A-A

Figure 5

Rotary Guide Assembly

Part Manipulator - To permit extreme contours to be trimmed using a basic two-axis numerical control system, a special part manipulator was designed and built to take full advantage of the laser as a cutting system (Figure 6). The z-axis height sensor is coupled with longitudinal and rotational axes of numerical control. The z-axis adaptive height sensor is used to automatically compensate for the surface height of a part as it is

Figure 6

Two-Axis Part Manipulator with Adaptive Z-Axis Motion

rotated at a preprogrammed rate (Figure 7). In effect, the unit can trim a part that is not a true body of rotation because of the vertical movement of the laser focus spot during rotation of the workpiece. The principal advantage of this system is its ability to handle large, severely contoured parts. The rotating tool makes it possible for the height sensing movement and normality errors to be a function of the radius of rotation rather than a flat plane.

IMPACT OF HIGHER POWER LEVELS

The effect of increased laser power (750 to 6000 watts) on the cutting process was evaluated at the United Technologies Research Laboratory with their 10.6-micron, continuous-wave carbon dioxide laser. Tests were conducted with and without assist gas. A schematic representation of the test arrangement is shown in Figure 8. The optimum focal spot size was 0.016 inch diameter with a f/4 focusing system. The materials evaluated were 4340 steel, Waspaloy, and titanium. Results are as follows:

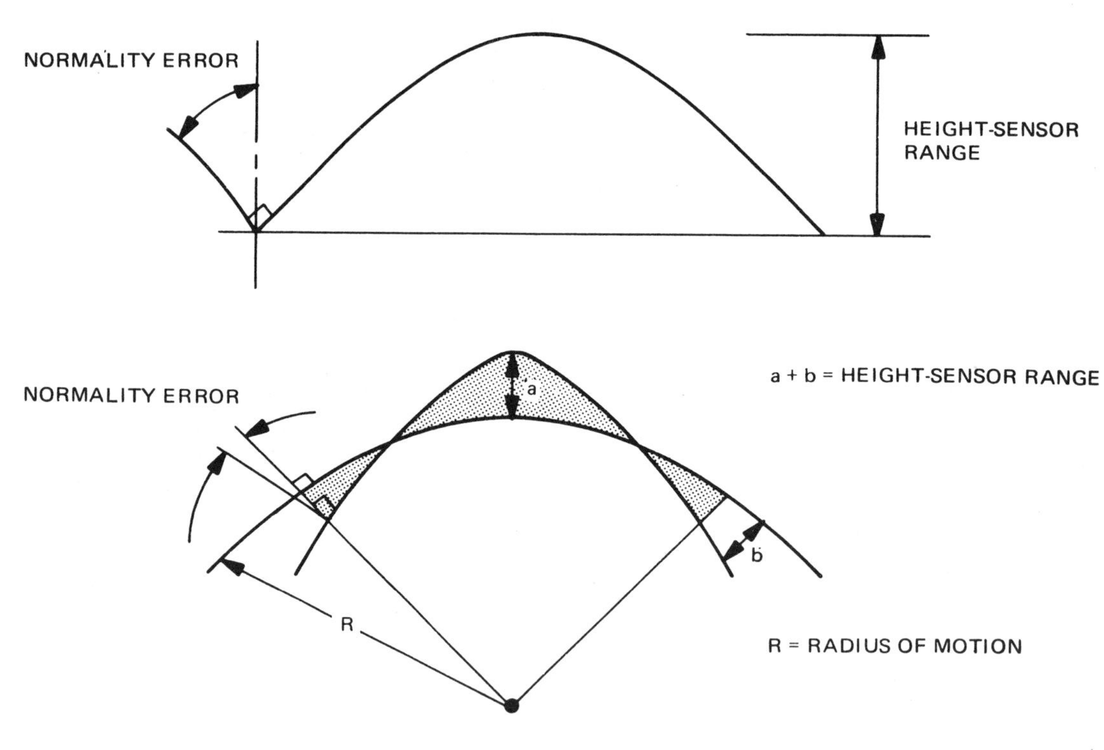

Figure 7

Principle of Rotating Tool

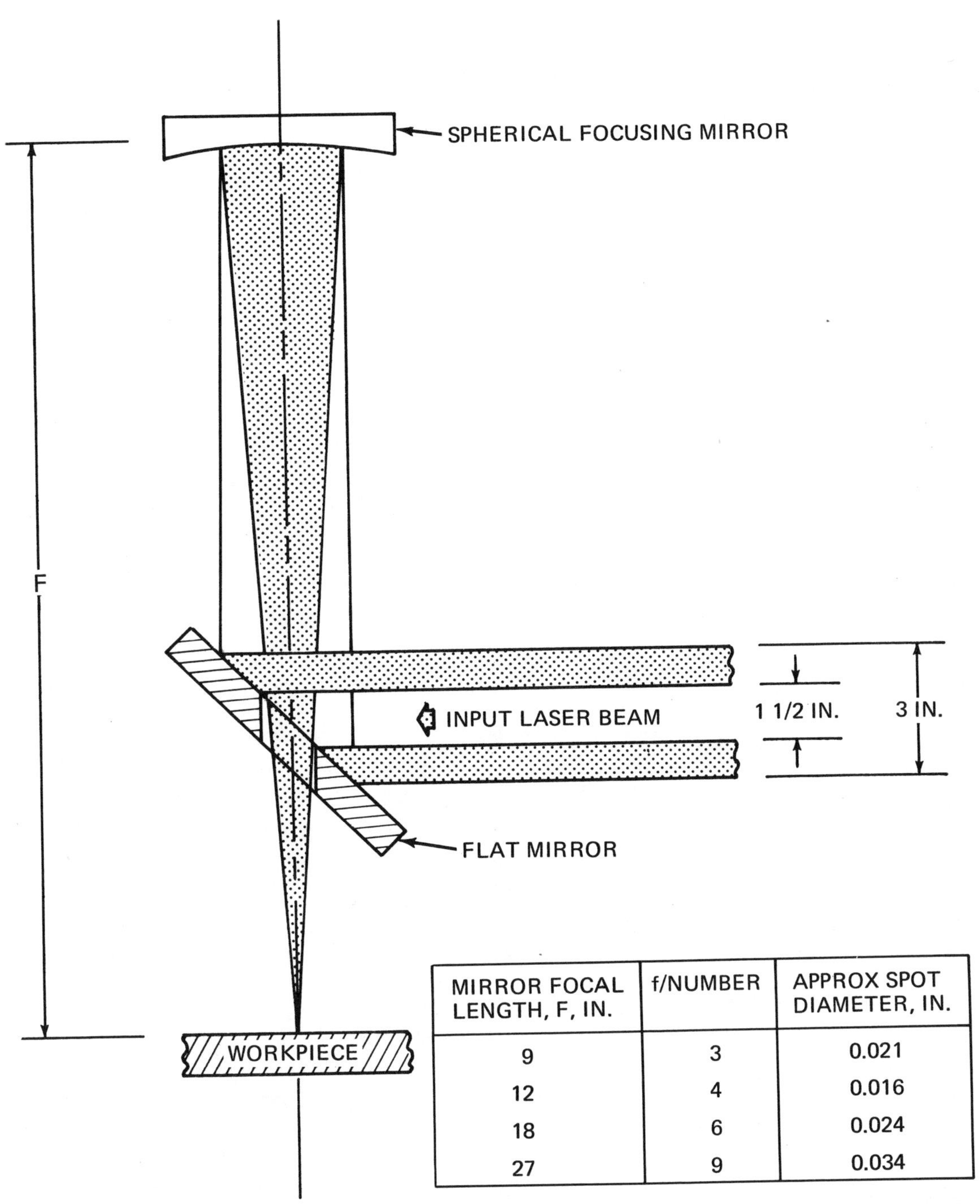

MIRROR FOCAL LENGTH, F, IN.	f/NUMBER	APPROX SPOT DIAMETER, IN.
9	3	0.021
12	4	0.016
18	6	0.024
27	9	0.034

Figure 8

United Technologies Test Arrangement

4340 Steel Alloy - Cutting performance (Table I) increased generally in proportion to laser power for both oxygen and inert-gas jet-assisted conditions. For oxygen-assisted cutting, it was found that the material could be severed at substantially higher speeds. In inert-gas-assisted cutting, optimum speeds were of the order of 50-60 percent of oxygen-assisted rates. A generally smoother edge was obtained.

Table I Laser Cutting Parameters for 4340 Steel Alloy

Thick, in.	Sample Best Cut Power, kW	Speed, ipm	Jet Pressure, psig	Approx. Kerf, in.		HAZ, in.	
				Top	Bottom	Top	Bottom
	Oxygen Jet Assisted						
0.125	0.75	50	75	0.032	0.052	0.008	0.013
0.125	1.50	70	120	0.035	0.041	0.0035	0.011
0.125	3.00	130	120	0.027	0.042	0.0075	0.025
0.125	6.00	200	200	0.021	0.024	0.004	0.014
0.125	1.50	45	200	0.040	0.043	0.0045	0.012
0.25	3.00	70	150	0.047	0.058	0.003	0.014
0.25	6.00	100	180	0.061	0.063	0.006	0.045
0.38	3.00	30	200	0.055	0.129	0.004	0.038
0.38	6.00	60	200	0.054	0.125	—	—
	Carbon Dioxide Jet Assisted						
0.125	1.50	50	100	0.016	0.014	0.0105	0.025
0.125	3.00	80	120	0.013	0.015	0.007	0.0255
0.125	6.00	100	250	0.020	0.016	0.0005	0.0315
0.25	3.00	15	200	0.021	0.030	0.024	0.033
0.25	6.00	30	200	0.025	0.025	0.022	0.033
0.38	6.00	18	350	0.025	.01-.14	0.015	0.057
	Nitrogen Jet Assisted						
0.125	3.00	80	200	0.020	0.022	0.005	0.014
	Air Jet Assisted						
0.125	3.00	80	200	0.024	0.014	0.0085	0.0205

Waspalloy - This material was readily cut at high speeds with generally acceptable cut quality being obtained (Table II). Since the material is resistant to oxidation, smaller gains in cutting speed were attained with oxygen than for the other two materials. As in the case of 4340 steel, the slag generated during inert gas-assisted cutting could readily be chipped from the edge leaving a relatively smooth cut surface.

Titanium Alloy - Oxygen-assisted cutting of titanium did not improve cut quality at higher power levels. The higher incident power density led to a violent reaction in the oxidation-prone titanium and to extremely rough cuts. For thick titanium sections, the primary cutting energy stems from the oxygen-jet assist such that cutting speed appeared to be essentially independent of power level. Inert-gas jet-assisted cutting of thin titanium sections gave a generally smooth cut surface with less apparent heat damage and easily removable adherent slag at the lower lip of the cut. A slightly serrated edge was achieved with carbon dioxide, nitrogen, and compressed air, the latter providing a substantial increase in cutting speed over that obtained with inert-gas-assist cutting.

Table II Laser Cutting Parameters for Waspaloy

Thick, in.	Sample Best Cut Power, kW	Speed, ipm	Jet Pressure, psig	Approx. Kerf, in.		HAZ, in.	
				Top	Bottom	top	Bottom
	Oxygen Jet Assisted						
0.03	0.75	400	100	0.010	0.010	0.005	0.0025
0.03	1.50	400	100	0.010	0.008	0.0005	0.004
0.03	3.00	600	100			0.0003	0.0015
0.03	6.00	720	175	0.022	0.018	Trace	0.0015
0.06	0.75	(1) 100	100	0.011	0.011	0.002	0.005
0.06	1.50	(2) 240	200	0.010	0.010	0.0025	0.003
0.06	3.00	350	100	0.018	0.018	0.004	0.0005
0.06	6.00	450	150	0.017	.025-.030	0.045	Trace
0.125	0.75	(1) 50	200	0.011	0.013	0.002	0.004
0.125	1.50	(2) 80	100	0.011	0.012	0.003	0.0055
0.125	3.00	140	120	0.019	0.023	0.0025	0.0055
0.125	6.00	270	150	0.017	.025-.030	0.0015	0.003
	Carbon Dioxide Jet Assisted						
0.03	0.75	(1) 140	100	0.012	0.008	0.0005	0.001
0.03	1.50	(1) 280	100	0.007	0.006	0.0005	0.0015
0.03	3.00	500	200	0.014	0.013	Trace	0.0005
0.03	6.00	650	250	0.016	0.015	0.0002	Trace
0.06	3.00	200	200	0.015	0.005	0.0005	0.002
0.06	6.00	350	200	0.023	0.022	0.0005	0.003
0.125	6.00	140	300	0.020	0.019	0.001	0.003

CoAx jet configuration with 0.06-in. diameter and 0.04-in standoff used in conjunction with f/4 optics except as noted.

(1) .04-in. jet diameter and f/3 optics

(2) .04-in. jet diameter, .06-in. standoff and f/3 optics

With helium assist, some of the irregularity in the cut surface was removed to give a relatively smooth edge. This may be due to the substantially higher jet velocity obtained with helium than with other gases. It appears that surface irregularities are influenced by appropriate control of the fluid dynamic characteristics of the jet (Table III).

Power levels of 10 and 16 kilowatts were evaluated at the AVCC Everett Research Laboratories using a 10.6-micron wavelength, carbon dioxide laser with a continuous wave output to 16 KW. The nominal focused beam diameter was 0.14 inch; the nozzle-to-workpiece stand-off ranged from 0.02 to 0.03 inch (Figure 9).

Table III Laser Cutting Parameters for Titanium Sheet

Thick, in.	Sample Best Cut Power, kW	Speed, ipm	Jet Pressure, psig	Approx. Kerf, in.		HAZ, in.	
				Top	Bottom	Top	Bottom
	Oxygen Jet Assisted						
0.25	0.75	60	60	0.060	0.075	0.0045	0.019
0.25	1.50	(1) 100	60	0.055	0.070	0.004	0.017
0.25	3.00	140	80	0.060	0.070	0.0035	0.016
0.25	6.00	180	150	0.102	.09-.16	0.018	0.037
0.75	1.50	(1) 50	100	0.062	0.123	0.0125	0.054
0.75	3.00	60	120	0.058	0.136	0.009	0.039
0.75	6.00	80	140	0.094	0.162	0.012	0.017
1.00	3.00	(2) 60	150	0.085	0.190	0.0115	0.046
2.00	6.00	40	200	0.118	0.500	—	—
	Carbon Dioxide Jet Assisted						
0.25	3.00	14	300	0.022	0.032	0.004	0.061
0.25	6.00	60	300	0.030	0.019	0.011	0.056
0.125	1.50	60	200	0.019	0.020	—	—
	Air Jet Assisted						
0.125	3.00	80	200	0.023	0.020	0.011	0.040
0.062	3.00	120	200	0.036	0.030	0.0115	0.021
	Nitrogen Jet Assisted						
0.125	3.00	80	200	0.022	0.022	0.007	0.023
0.062	3.00	120	200	0.016	0.015	0.006	0.0075
0.125	3.00	70	20	0.012	0.012	0.006	0.018

(1) .054-in. jet diameter and f/6 optics
(2) .050-in. jet diameter

For laser cutting of titanium, in general, reactive gases such as oxygen are not compatible with high-power lasers and result in uncontrollable cut quality. Although the use of oxygen to cut 3/4-inch-thick 4340 steel alloy resulted in parallel cut faces, the kerf was much wider than that observed in 2-inch-thick titanium laser cut with helium. A summary chart (Table IV) identifies recommended assist gases for high-power laser cutting of titanium and steel alloys. A summary of best-quality cuts is shown in Table V.

Table IV Cutting Gas Summary

Cutting Gas	Titanium	Carbon Steel	Remarks
Helium (100%)	Best	Good (Costly)	
Helium/10% Argon	Best	Not Evaluated	Air works for carbon steel and is less costly
Air	Not Recommended	Best	
Carbon Dioxide	Not Recommended	Not Evaluated	Air works for carbon steel and is less costly
Nitrogen	Not Recommended	Not Evaluated	Air works for carbon steel and is less costly
Oxygen	Not Recommended	Not Recommended	Severe undercut

Table V Summary of Best-Quality Cuts in 10-16 Kilowatt Range

Material	Thickness, in.	Power, kw	Gas	Feedrate, in./min	Kerf, in.		HAZ, in.	
					Top	Bot	Top	Bot
Ti-6Al-4V Titanium	2	15	He	5	—	—	0.028	0.165
↓	1	15	75% He/25% A	15	—	—	—	—
↓	1	15	He	15	0.093	0.046	0.036	0.090
↓	3/4	10	90% He/10% A	19	0.082	0.029	0.040	0.180
↓	3/4	10	He	19	0.080	0.037	0.019	0.073
↓	1/4	10	He	80	0.061	0.019	0.015	0.395
↓	3/4	10	He	15	0.067	0.036	0.034	0.105
4340 Steel	3/4	10	He	10	0.090	0.054	0.044	0.099
↓	3/4	15	75% He/25% A	15	—	—	—	—
↓	3/4	11.5	O_2	40	—	—	—	—
↓	3/4	10	Air	9.4	—	—	—	—
↓	3/4	15	He	15	0.090	0.037	0.032	0.097
↓	3/4	10	O_2	45	—	—	—	—

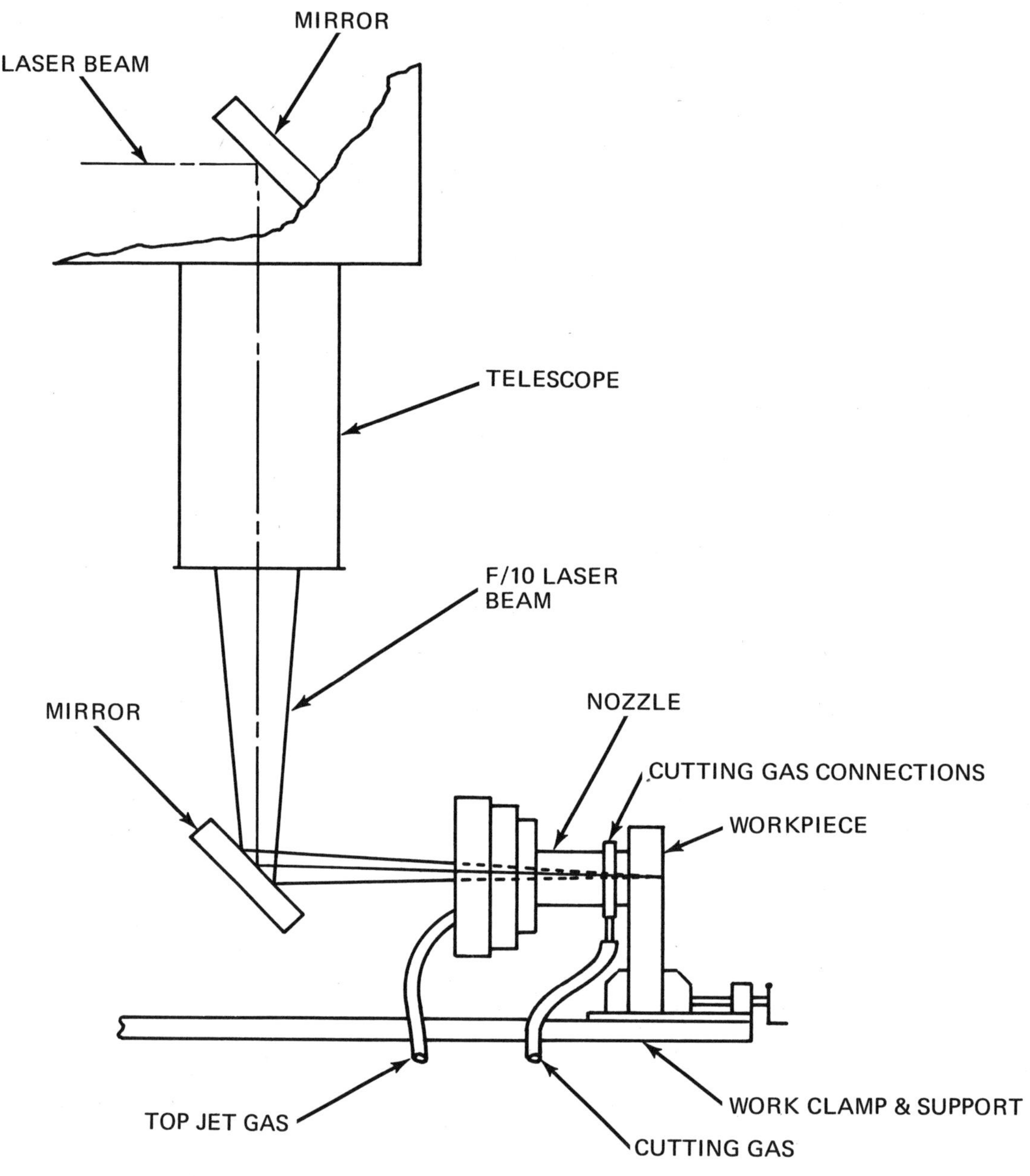

Figure 9

AVCO Test Arrangement

ECONOMIC EVALUATION

Two basic areas are involved in the cost of a laser cutting system. These are the capital investment to establish a production cutting facility and the operating costs of the equipment. Laser system costs for a broad range of output power levels are shown in Figure 10. These costs are based upon current state-of-the-art laser beam generating systems and do not include beam manipulation or position control systems. The beam may be delivered to the workpiece by either of two means: movable laser or movable optics. Although the movable laser is a lower-cost approach ($100,000 versus $300,000-400,000 for the movable optics), movable optics lend themselves to high-speed cutting systems (low inertia). It

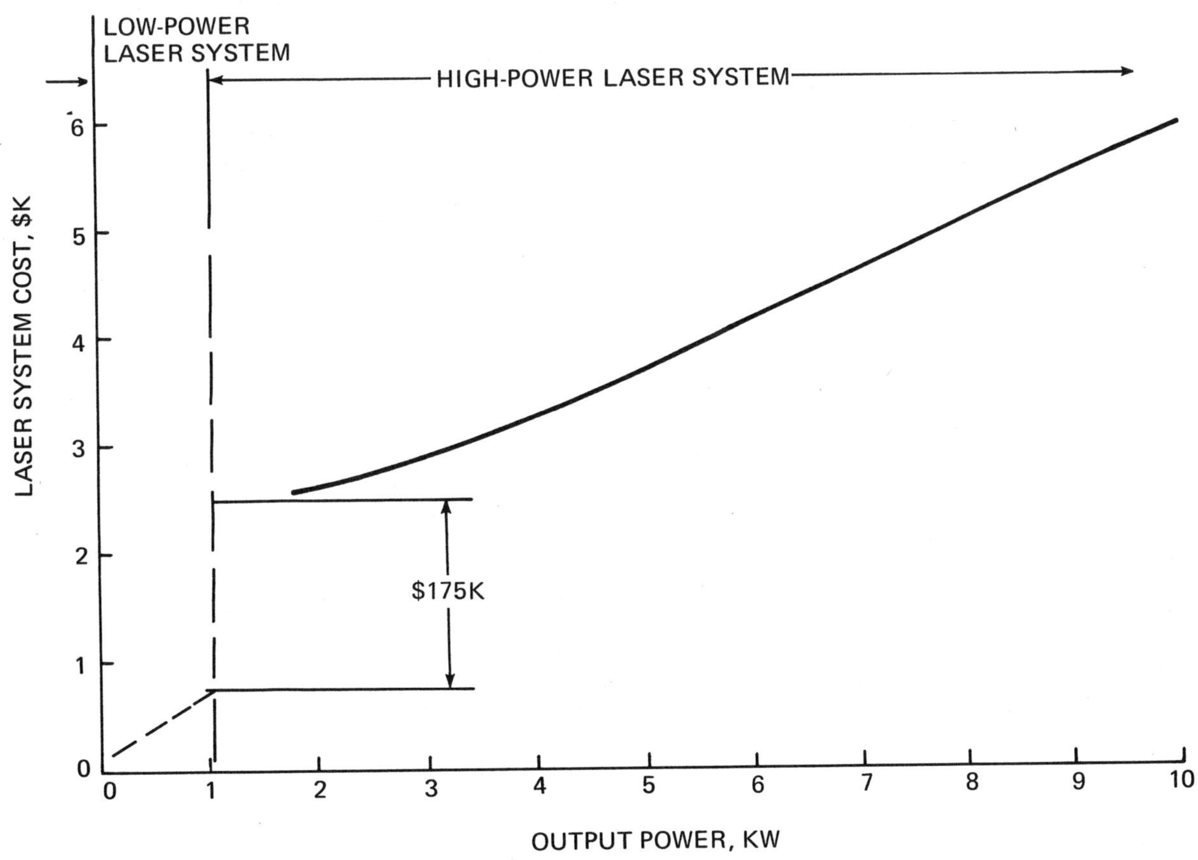

Figure 10

Laser System Cost-Output Power Relationship

should be noted that there are no currently available, movable lasers above the 500-watt category. A compilation of laser system and beam handling apparatus costs yields the total system cost-laser power relationship shown in Figure 11. These costs represent those of a total cutting system exclusive of installation.

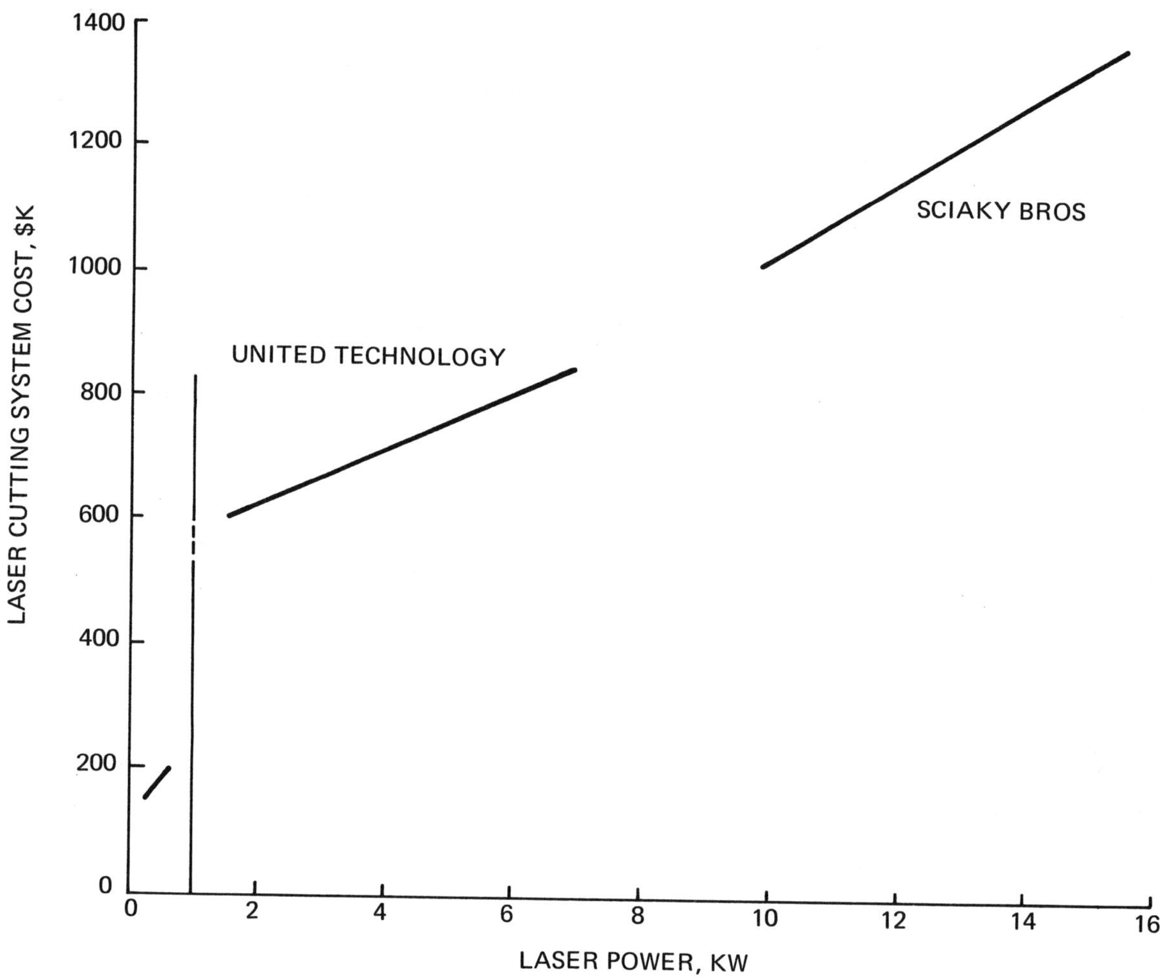

Figure 11

Laser Cutting System Cost-Laser Power Relationship

The cost of generating the laser beam includes input power and laser gas. Projected hourly laser operating costs based upon a 100-percent duty cycle are given in Figure 12 using an electrical power charge of \$0.035 per kilowatt hour. A complete economic analysis would also require a comparison of assist-gas costs and post-processing cost differentials between the particular processes being compared.

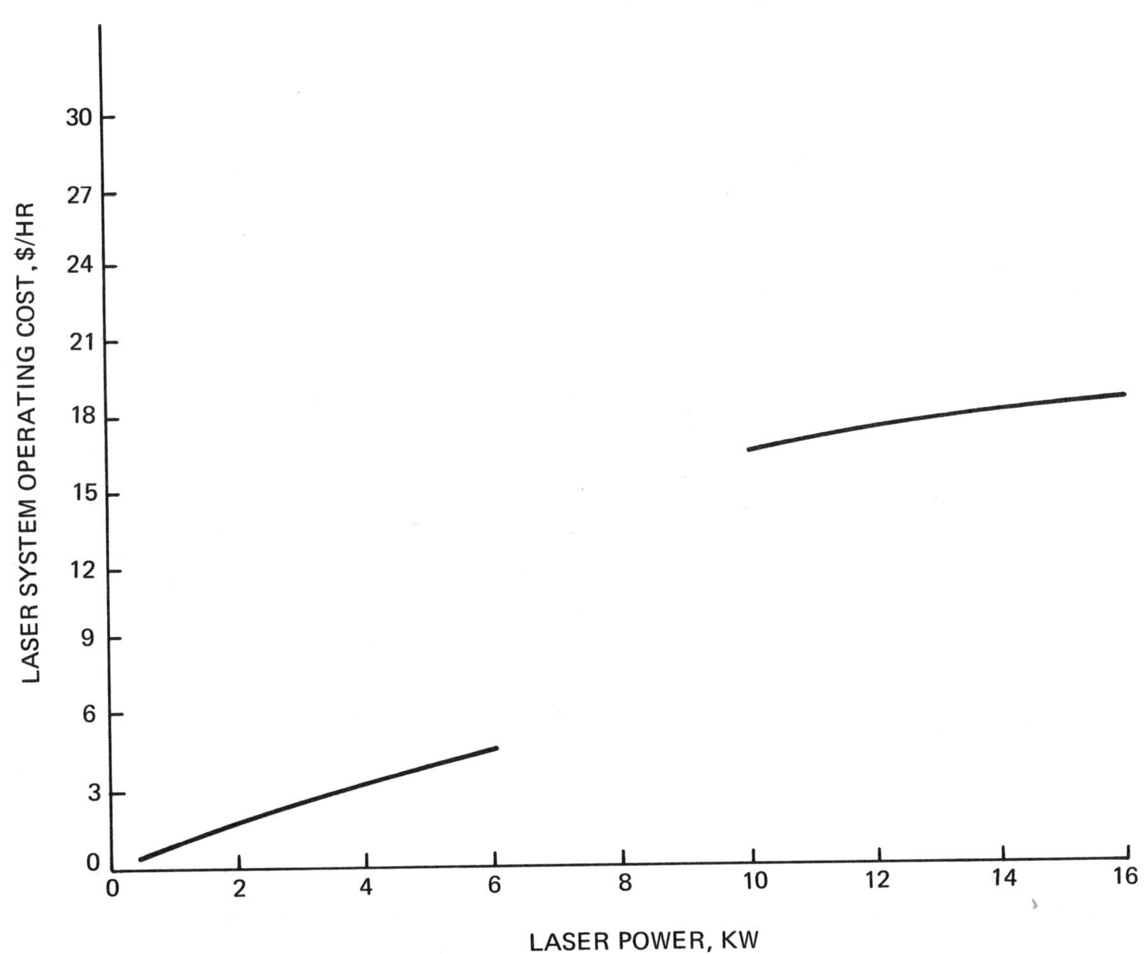

Figure 12

Effect of Laser Power on Laser System Operating Cost

PRODUCTION APPLICATIONS

The laser cutting process at Grumman has been primarily applied to titanium alloy aircraft parts. In each of the applications, laser cutting was used as a rough trimming operation with a post-process sanding or routing on all parts. The parts selected for discussion represent the overall spectrum of part sizes and shapes cut using the laser. A number of the formed parts trimmed involved the use of some unique innovations for cutting using a basic two-axis numerical control system. Tooling for these parts was generally minimal consisting of the numerical control tape and a light-duty positioning arrangement. The actual positioning holding forces are minimum, since the laser does not impose cutting forces on the part itself. In each of the examples listed, the laser reduced the rough cutting time by over 50% compared with that for conventional cutting processes.

Stabilizer Splice Plate - Cost evaluations showed that laser cutting would be more cost-effective than the techniques then being used in production to cut the titanium splice plates for the F-14A horizontal stabilizer. These involved chemically blanking 5 x 11-foot, Ti-6Al-6V-2Sn titanium alloy plates previously surface ground to a uniform thickness of 0.350 inch. Chemical blanking was accomplished by masking the plate, scribing the tracks, peeling off the maskant from the tracks, and chemically etching the plate. The parts were removed from the chem-milled plate by chiseling away the remaining metal in the tracks. The sharp, ragged edges were smoothed by grinding. Two splice plates were obtained from each 5 x 11-foot blank (Figure 13).

Laser cutting of the splice plates (Figure 14) is much simpler compared to the chemical milling/chiseling/grinding operation. It involves numerically controlled laser cutting followed by slight deburring with a disc sander to smooth the edges. The ease and rapidity with which the splice plates could be laser cut, after the blanks had been positioned on the worktable, led to an increase in the size of the blank from 60 x 132 inches to 79 x 235 inches and an increase in the number of splice plates cut per blank from two to six. The old and new cutting configurations are shown in Figure 13. The actual cutting time was about 5 minutes per plate at a 60 inch-per-minute feed rate. This gave an overall savings of 17.6 man hours and a $1350 reduction in material cost per aircraft.

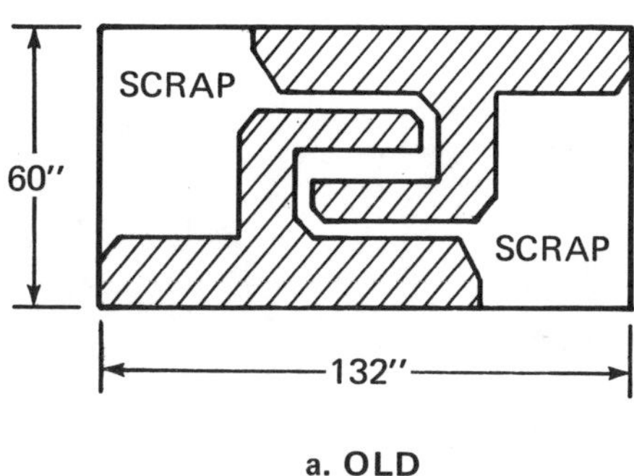

a. OLD

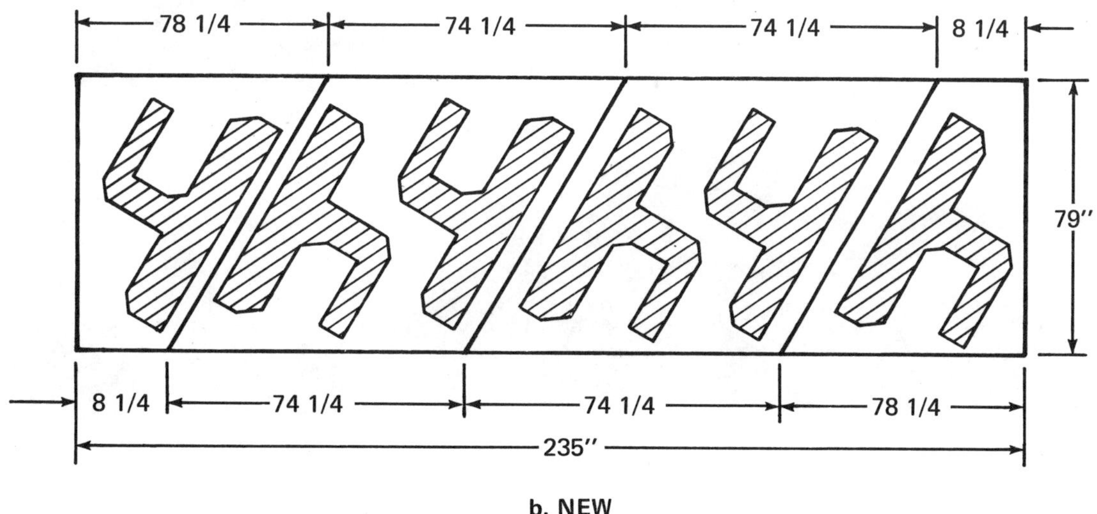

b. NEW

Figure 13

Laser Cutting Configurations for F-14A Titanium Stabilizer Splice Plate

F-14A Wing Stringers - The stringers, which range in length from 8 to 21 feet, are formed from material tapered in thickness from 0.170 to 0.050 inch. The stringers were trimmed originally by bandsawing as close as possible to the required shape, routing and then belt-sanding to the final trim line.

Figure 14

Laser-Cut F-14A Titanium Stabilizer Splice Plates

Belt sanding also removed any heat-affected material remaining from the bandsawing and routing operations. About 120 minutes were required to fabricate each stringer by this method. Laser trimming (Figure 15) reduced the fabrication time per stringer to about 20 minutes; this also included loading and unloading of the holding fixture. Actual laser cutting time was only five minutes.

Because these formed titanium stringers are not completely flat, a height-sensing range of about one inch is required to assure a good quality cut through the length of the part.

Figure 15

Laser Trimming of F-14A Wing Stringers

Stiffener - This is a flat, 1/16-inch-thick, 2 x 30 inch steel part (Figure 16). The tooling used consisted of a series of standoffs with three small permanent magnets to hold the part in position during the laser cutting process. The actual cutting time was 1 1/2 minutes versus 20 minutes for bandsaw cutting. This illustrates that even without the very high cutting rates of titanium the ability to use a laser with minimal tooling and the use of automated control can produce significant benefits.

Figure 16

Laser Cutting of Steel Stiffener

Nacelle Skin - This is a 0.090 x 22 x 80-inch titanium alloy part with several cutouts and a minimum radius of curvature of 50 inches (Figure 17). The moderate contour of this part is typical of aircraft skins including fuselage areas. The previous method for trimming this part was chemical etching. A light frame tool was used to hold the part in its proper contour during laser cutting. An extended Z-axis adaptive height sensor was developed to provide for a range of up to 10 inches variation in contour. Laser trimming reduced the trimming cost of this part by 58%.

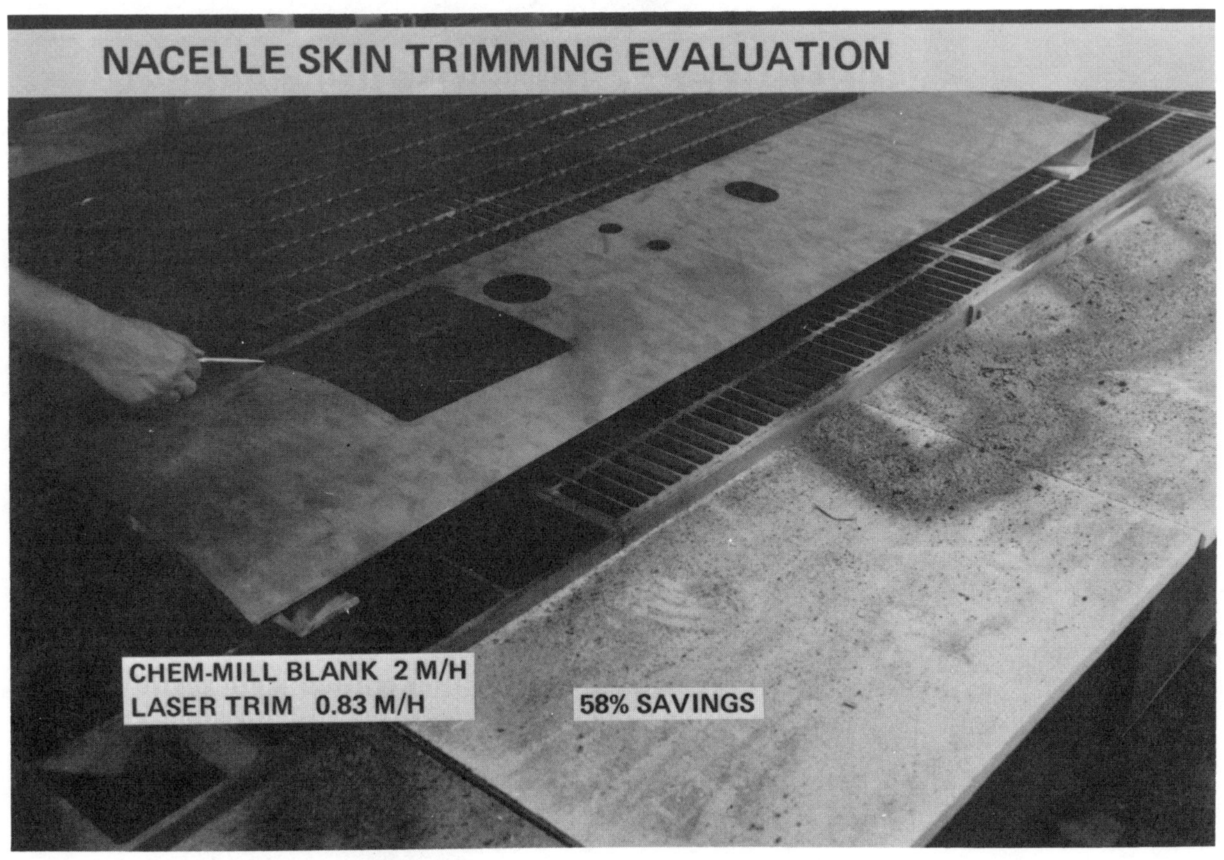

Figure 17

Laser-Cut F-14A Nacelle Skin

Nacelle Skin Severely Contoured - A 0.090-inch-thick Ti-6Al-6V-2Sn titanium alloy skin was cut using the special part mainpulator described previously. Total laser cutting time was three minutes and fifteen seconds per skin. The rough trimming cost savings realized by using the laser in place of chemical blanking was about 54%.

Small Parts - The use of the laser for trimming small parts without numerical control was demonstrated using the rotary guide described previously. Several parts and their associated savings are shown in Figures 18 and 19.

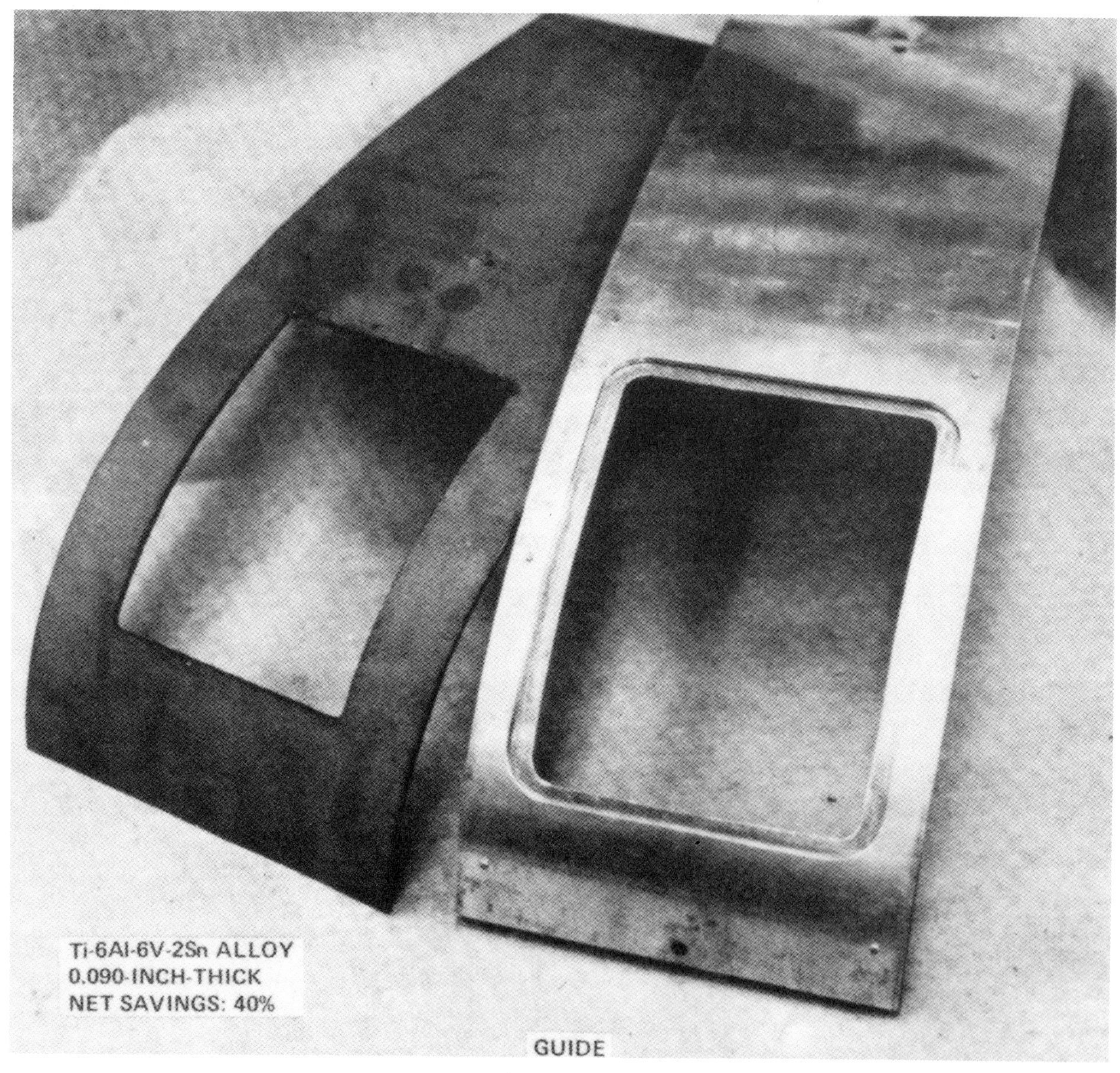

Figure 18

Titanium Part Laser - Trimmed with Manual Rotary Guide

Figure 19

Steel Parts Laser-Trimmed with Manual Rotary Guide

REFERENCES

1. Meyer, M., "Laser Cutting Manufacturing Methods", Air Force Technical Report No. AFML-TR-73-138, May 1973

2. Marx, W. "Manufacturing Methods For Multi-Axis Laser Cutting", Air Force Technical Report No. AFML-TR-75-169, November 1975

3. Trink, S., "Laser Cutting Of Aerospace Materials" SME Paper No. EM 73-214

CHAPTER 7

ELECTRON BEAM MACHINING

Reprinted from: Tooling and Production, March 1976

Electron-beam machining—more than joining

DRILLING is the key word, but the action includes much more. Beginning with electron-beam welding (EBW), researchers have advanced the art of *electron bombardment* to include perforating, drilling, milling, engraving, metal hardening, fusion, transformation and sublimation.

The success of the new applications of electron-beam machining (EBM) result from the development of an improved electron-beam generator that provides more precise beam positioning and makes possible highly accurate machine tools. The machines are available in chamber and continuous types, or in the cycle type for high-volume production. They are made by Steigerwald Strahltechnik GMBH, Puchheim, Germany, and are sold in the US by Farrel Co. Div., USM Corp., Ansonia, CT.

Applications include perforation and polymerization of plastic. According to Dr. K. H. Steigerwald, "By using low power densities, irradiation of monomeric paint systems induces chemical processes such as polymerization within fractions of a second. Besides beam curing, EBM can produce high-resolution integrated circuits with the electro-resist technique.

"By focusing the beam to a small spot and increasing power density, we can perforate plastic materials for filters. And one of the most important uses is the perforation of synthetic leather for shoes—to make it permeable for effective transpiration.

"Based on electron bombardment, we developed a steel-hardening technique that gives excellent hardness within defined areas. With power densities in the range of 10^5 to 10^6 W/cm^2, practically all metals can be melted. From this fact, vacuum melting of pure metals, fusion treatment of workpieces and the wide field of electron-beam welding derive.

"If the energy input within a small volume is high enough to evaporate the material, it is possible to perforate, drill and engrave even the hardest materials at high speed. Maximum power densities (10^9 W/cm^2) are required by sublimation processes."

Machine design

As in EBW, a triode generates the beam. A heated cathode emits electrons, thus forming a space-charge cloud, and a voltage of 100 to 150 kV between cathode and anode impacts high velocity to the electrons. A bias electrode, negatively charged relative to the cathode, switches the beam current on and off, thus forming the pulses required for all drilling operations. It also controls the intensity of the beam. *An electro-magnetic lens focuses the divergent electron beam to a power density at least two orders of magnitude higher than used for welding applications.* Magnetic coils deflect the beam in any desired direction. To avoid dispersion of the beam by collision of the electrons with gas molecules, the whole system of beam generator and working chamber is evacuated —the acceleration space to a pressure less than 10^{-4} mmHg, and the working chamber to about 10^{-2} mmHg.

The system offers high stability of the electrical parameters as well as controllability by a computer.

Farrel-Steigerwald drilling machine, one of several configurations of the new big guns for electron bombardment. These can drill holes from 25 μm dia in 20 μm thick material up to 1 mm dia in 5 mm thick material. In welding models, beam powers range from 100 W at 1.5 kV to 60 kW at 150 kV. Weld depths range from 0.040" to 7.87". In multipulse operation, the EBM equipment can drill or mill noncircular holes, including triangle and I shapes.

The average beam power is 1 kW with overload up to 15 kW within a single pulse. The focal length depends on the speed of the electrons, so stabilization is vital. The pulse width can be varied continuously between 10 μsec and 20 msec, and the pulse frequency ranges up to 10 kHz. The machine can be controlled manually or by a digital computer. For these studies, a special computer control was incorporated in the equipment.

The electron beam of charged particles has very low mass and very high velocity, thus offering ideal conditions for the application of numerical control. The low-inertia beam, controlled by electromagnetic fields, allows extremely fast processing of intricate machining geometries. Simultaneous control of beam and workpiece movement is provided, and the beam can accurately strike a moving target.

The on-line computer controls workpiece movement, beam deflection and beam focus. *This combination allows three-dimensional movement of the focal point in combination with a defined operating speed on the workpiece.* Furthermore, the computer controls the beam current as a function of workpiece position and time, and also triggers the energy pulses. The advantage of this system is that the actual physical positions and electron-beam parameters are continuously recorded in the computer and compared with the required positions and parameters that have previously been stored in the computer memory. The error signal is then used to correct any deviations.

The workpiece moves continuously under the beam and causes the computer to trigger the electron-beam pulse when hole position is reached. During the pulse width, the beam follows the motion of the workpiece to impact on the exact spot.

EB drilling

If electrons collide with solid material, their kinetic energy immediately translates into thermal energy. What happens within the workpiece depends on the electron-beam parameters such as total power, power density and duration of impact. It also depends on the thermal properties of the target such as heat capacity, level of melting and vaporizing points, heat conductivity, and heat of fusion and vaporization. *In short, the material can be heated, melted or vaporized.*

For holemaking, the material is either evaporated or is only melted and the liquid phase taken away by additional actions such as centrifugal forces. In general, a combination of fusion and evaporation is used so that the vapor pressure ejects the liquid material. As in EBW, the pulsed beam also creates a capillary and a cylindrical fusion zone. Because of the high power density in such a beam, the vapor pressure in the capillary increases so much that the material is ejected in a sudden burst.

To save energy and thus optimize the economy of the process, the minimum evaporation rate is aimed for. Roughly 75 percent of the energy of impulse is necessary to melt the volume to be removed. The remaining 25 percent of the energy evaporates about 5 percent of the volume. The pulse interval, and thus the time of impact of the beam on the workpiece, is between 10 μsec and 10 msec. The main task of the process control is to choose suitable beam parameters for proper molten volume, temperature, heat-influenced zone etc.

The main control parameters for shaping the hole are the pulse width for the depth of the hole, the beam current for the diameter of the hole, the power distribution within the beam and the position of the focus with respect to the workpiece. These parameters can be combined in many ways, and each can be varied with respect to time. The resulting range of possible hole shapes is wide, and multipulse drilling with pulse frequencies between 50 and 1000 Hz can serve where a single pulse per hole can't do the job. Sequences of hundreds to thousands of pulses can machine most any shape, and reproducibility of the machining operations is ensured by a suitable computer control.

Other techniques

Sublimation is based on a quite different mechanism of material removal. By increasing the power density, a direct transition of the solid into the gaseous phase is achieved. To avoid a molten pool resulting from heat conductivity, extremely short impact times are advantageous. This can be obtained by high-frequency pulsing of the beam or by a fast relative movement between workpiece and beam.

If the efficiencies of single-pulse drilling (perforation), multipulse drilling and sublimation are compared, the removal rates of volume per unit of time have the ratios of 1000:10:1. The result, however, has its reason partly in the physics of the removing process and partly in the equipment.

The present state of the EBM art allows removal rates of up to 40 mm^3/sec with the single-pulse technique. This is some orders of magnitude faster than EDM, ECM or laser-beam machining (LBM). The reproducibility and achievable tolerances are within ±5 percent of the nominal value, and the surface roughness is about 5 μm, i.e., the quality is similar to EDM and ECM and much better than LBM. All this according to Dr. Steigerwald, who expects EBM performance to be even further improved.

Holes can be EBM drilled at angles from 20 to 90 degrees, as shown by this gas-turbine mixer plate of titanium alloy. Hole diameter is 0.3 mm. In a plastics application, one machine drills 5000 holes a second to perforate artificial shoe leather to let it breath, adding to user comfort. The process can drill deep holes with diameter-to-depth ratios of 1:10. Normal EBM holes are round with trumpet-shaped entrance, but elliptical cross sections are possible, as are tapered holes.

Reprinted from: Machine and Tool Blue Book, January 1976

Farrel-Steigerwald electron beam drilling machine. In this setup, sheet to be drilled is curved around drum shown protruding through opened chamber door at right-hand side of machine. Minicomputer console (right foreground) controls feed and revolution of drum in conjunction with operation of EB gun to yield required hole size and pattern.

Electron beam machining aims at broader applications

The electron beam is a versatile thermal tool. Up to now only electron beam welding has found its way into industrial applications with approximately 1000 EB welding machines being installed in this country

But the state of the art of EB techniques today has been demonstrated recently by the Farrel Co., a division of USM Corp., in Ansonia, Conn. Perforation of plastic materials is done under low power densities. Electron bombardment for steel hardening results in controlled surface hardness in defined areas.

At higher power densities vacuum melting of pure metals, fusion treatment of workpieces and a broad field of electron beam welding can be handled. If energy input within a small volume is high enough to vaporize the work material it is possible to perforate, drill and engrave even the hardest materials at high speed.

Under an agreement with Steigerwald Strahltechnik of Munich, West Germany, Farrel has acquired exclusive rights in the U.S. and Canada to manufacture and market electron beam products. The heart of the system is the Steigerwald electron beam generator which provides precise beam positioning with either manual or computer control and makes possible highly accurate machine tools for a variety of applications.

Standard Steigerwald EB generators and guns are shipped to Farrel which develops complete process lines, installations and start-up procedures based on customer requirements.

Three different machine configurations were shown in operation. A chamber type partial-vacuum machine was used for welding, fusion treatment, hardening and engraving. A precisely limited area of the workpiece is heated by the focused electron beam within milliseconds to the required temperature. Depending upon the nature of the intended structure change, the material is treated either

Here is a good example of the results of EB drilling on a production basis. This is a high-temperature part for a helicopter gas turbine engine. Note not only the number and cleanliness of the holes, but also the angle at which they were drilled as indicated by the two pins.

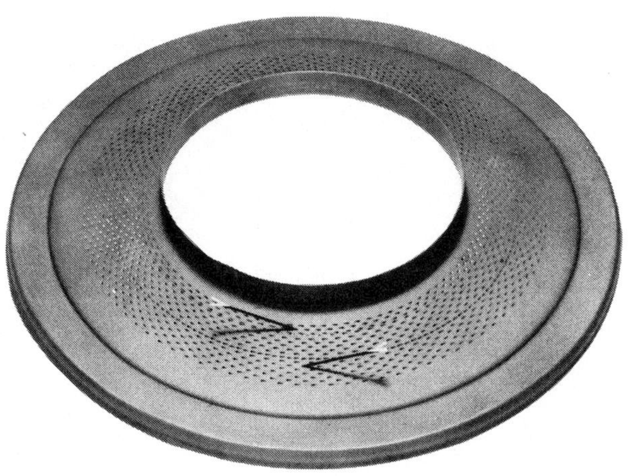

High-production cycle type welding machine. It is of rotary table indexing design and is intended basically for parts requiring circular welds. The partial vacuum needed for welding can be pumped down in 10 seconds; with operational equipment, 6 seconds.

by a remelting cycle or by transformation in the solid state. Duration of the temperature cycle is minimal. This technology provides many ways to improve design and manufacture of heavy-duty workpieces.

A second, computer controlled chamber type partial-vacuum machine used electron beam energy for drilling and perforating small and precise holes. This type of equipment is being used for perforating breathing holes in synthetic leather for shoe tops at the rate of 5,000 holes per second.

The angle at which the beam can hit the target can be as close as 20 degrees from the workpiece surface. This capability has led to the EB drilling of turbine blades, combustion chamber rings, mixer plates and other gas turbine parts exposed to high temperatures. A more recent development in blade cooling is called transpiration cooling which demands up to 30,000 holes per blade. This is economically possible with high-speed EB drilling.

A wide range of materials can be welded by electron beam energy on a production basis in the partial-vacuum cycle welder designed to handle medium and large batches of parts. It is a general purpose, multiple-station rotary-table indexing machine for EB joining of parts requiring basically circular welds. The main frame is sized to permit the addition of secondary automatic loading, unloading or inspection units.

Operation of the machine is quick. Total cycle time, including pump down (but excluding weld time which takes but a few seconds depending on the parts), is a maximum of 15.5 seconds. ●●●

INDEX